Advances in Techno-Humanities

This book is a pioneering attempt to explore the relationships between technology and the humanities through case studies and specific contexts in the areas of language, theatre, literature, translation, philosophy, music, home designations, learning environment, and artificial intelligence.

Written by scholars and specialists across various fields, the chapters explore the emerging field of techno-humanities. This book examines the development of language and society by means of Big Data, how technology is integrated into the theatres of Hong Kong and the ensuing results of such integration. The authors also highlight how technology is able to analyse, understand, and visualise literary works and to bring drastic changes to translation in the past seven decades. Long-standing philosophical issues are re-examined, linkages between technology and theoretical concepts are illuminated, and the emotional aspects of computational applications are investigated. This book also delves into insightful case studies such as providing suggestions to train novice translators through corpus-assisted translation teaching, analysing patterns of housing names, and discovering a new online method to acknowledge acquisition through authentic learning experiences. Overall, this book serves as a point of departure for us to go deeper into the role of technology in transforming the humanities in this digital age.

This is a useful read for students and scholars interested in learning more about the cross section between humanities and technology.

Mak Kin-wah is President of Caritas Institute of Higher Education, which strongly supports Techno-Humanities teaching and hosts a research centre dedicated to this purpose. He holds the degrees of PhD and MPhil from Cambridge University, an MBA from City University London, and a Bachelor of Engineering from the University of Western Australia.

Routledge Studies in Translation Technology and
Techno-Humanities

This cutting-edge research series examines translation technology and techno-humanities and explores the relationships between human beings and machines in translating the written and spoken word and the connection between technology and the humanities. The series welcomes authored monographs and edited collections.

Series Editor: *Chan Sin-wai*

The Future of Translation Technology
Towards a World without Babel
Chan Sin-wai

The Human Factor in Machine Translation
Edited by Chan Sin-wai

Controlled Document Authoring in a Machine Translation Age
Rei Miyata

Analysing English–Arabic Machine Translation
Google Translate, Microsoft Translator and Sakhr
Zakaryia Almahasees

Advances in Techno-Humanities
Case Studies from Culture, Philosophy and the Arts
Mak Kin-wah

For more information on this series, please visit www.routledge.com/Routledge-Studies-in-Translation-Technology/book-series/RSITT

Advances in Techno-Humanities
Case Studies from Culture, Philosophy and the Arts

Edited by Mak Kin-wah

LONDON AND NEW YORK

First published 2024
by Routledge
4 Park Square, Milton Park, Abingdon, Oxon, OX14 4RN

and by Routledge
605 Third Avenue, New York, NY 10158

Routledge is an imprint of the Taylor & Francis Group, an informa business

© 2024 selection and editorial matter, Mak Kin-wah; individual chapters, the contributors

The right of Mak Kin-wah to be identified as the author of the editorial material, and of the authors for their individual chapters, has been asserted in accordance with sections 77 and 78 of the Copyright, Designs and Patents Act 1988.

All rights reserved. No part of this book may be reprinted or reproduced or utilised in any form or by any electronic, mechanical, or other means, now known or hereafter invented, including photocopying and recording, or in any information storage or retrieval system, without permission in writing from the publishers.

Trademark notice: Product or corporate names may be trademarks or registered trademarks, and are used only for identification and explanation without intent to infringe.

British Library Cataloguing-in-Publication Data
A catalogue record for this book is available from the British Library

ISBN: 9781032453255 (hbk)
ISBN: 9781032453323 (pbk)
ISBN: 9781003376491 (ebk)

DOI: 10.4324/9781003376491

Typeset in Galliard
by Newgen Publishing UK

Contents

List of Figures	*vii*
List of Tables	*ix*
More about the Editor	*xi*
List of Contributors	*xii*

	Introduction MAK KIN-WAH	1
1	Towards the Identification and Tracking of Salient Traits and Their Developments in Chinese Society via Big Data BENJAMIN K. TSOU, KELLY MAK, AND KENNY MOK	4
2	Techno-humanities: Some Trends of the Portrayal of Science in Art on the Hong Kong Stage THOMAS LUK YUN-TONG	19
3	Densities and Fugitive Vectors GRANT HAMILTON	28
4	Revisiting the Future of Translation Technology CHAN SIN-WAI	44
5	The Idea of Techno-philosophy and Philosophy-aided Technology, with Social Networking as an Example YING KOON KAU	66
6	Corpus-assisted Translation Learning: Attitudes and Perceptions of Novice Translation Students LIU JIANWEN, SU YANFANG, AND LIU KANGLONG	76

Contents

7   What Is an "Ideal" Home? A Multimodal Discourse
    Analysis of the Housing Names and TV Advertisements
    in Hong Kong   93
    LAM YEE MAN, LAM SHU YAN, AND NG KWAN-KWAN

8   A Conceptual Framework for Integrated Immersive
    Learning with Metaverse and Student-generated Media   113
    WONG PUI YUN, WONG WAI CHUNG, AND SHEN JIANDONG

9   Problems of Exacerbation to Dasein in the Modern
    Technological World by Use of the Early Heidegger's
    Theories: Readiness-to-hand and Presence-at-hand   127
    LAU HOK-YIN

10 Ethically Speaking: Opportunities and Risks of AI
    Chatbots Showing Empathy to Customers during
    Service Encounters   143
    YEUNG WING LOK

    Index   *154*

Figures

1.1	Sentiment Polarity of the QIE 今非昔比 "Not like what it used to be before" across Four Pan-Chinese Communities in 2005–2013	8
1.2	Overall Northbound and Southbound Lexical Transfers between Mainland China and Hong Kong (1997–2011)	11
3.1	Feature co-occurrence network of *Waiting for the Barbarians*	32
3.2	Section network of Coetzee's *Waiting for the Barbarians*	36
3.3	Graph showing the cumulative sum of the emotional valence of *Waiting for the Barbarians* and the position of sections 53, 99, and 100	37
3.4	Graph showing the cumulative sum of the emotional valence of *Waiting for the Barbarians* and the change in frequency of negative emotional language that occurs at section 53	38
3.5	Graph showing the distribution of topic 119 (warrant officer) in *Waiting for the Barbarians*	39
3.6	Graph showing the distribution of topic 10 (day_day) in *Waiting for the Barbarians*	40
4.1	Tools in translation technology	45
4.2	Systems in translation technology	46
4.3	Direct mapping procedure	49
4.4	Example-based machine translation procedure	50
4.5	Syntactical transfer in RBMT	51
4.6	Gap-filling in RBMT	51
4.7	Divisions of statistical machine translation	52
4.8	Three stages of artificial intelligence	56
4.9	Reusability in computer-aided translation	56
4.10	Periods of development of computer-aided translation	56
4.11	Global development of computer-aided translation, 2019	57
4.12	Speech processing in a speech translation system	61
4.13	Major trends of computer-aided translation	63
6.1	Homepage of the parallel corpus	80
7.1	Tsuen Wan Garden, Google map street view	100

viii *List of Figures*

7.2	Tsuen Wan Garden, Google map	100
7.3	Dragonfair Garden, Google map street view	101
7.4	Dragonfair Graden, Google map	102
8.1	Social-integrated metaverse learning conceptual model	115
8.2	Screenshots of the metaverse learning application, Classlet	119
8.3	Screenshots of Soqqle video application with game-based learning	120
8.4	Screenshots of Soqqle video application with reflective learning	121
8.5	Screenshots of Soqqle video application with evidence-based learning	122
9.1	Justification of the problems to Dasein's existence posed by the problems related to technology	136

Tables

1.1	Positive and Negative Connotations of QIE 今非昔比 "not like what it used to be before" Based on LIVAC	6
1.2	Reference to Chinese National Leaders in the HK Media (1995–2002)	9
1.3	Attributes of Proper Names in Hong Kong (Pre- and Post-1997)	10
1.4	Lexical Transfer Northbound and Southbound (1997–2011)	11
1.5	Percentage Distribution of Terms with the Two Gender Headwords in Beijing, Hong Kong and Taipei (1996–2000 and 2011–2015)	14
1.6	Connotations of Top Frequency Terms with the Two Gender Headwords	15
1.7	Examples of Top Frequency Terms with the Two Gender Headwords in LIVAC	16
3.1	Frequency of the 15 Most Common Words in *Waiting for the Barbarians*	30
3.2	Top TF-IDF Terms for *Waiting for the Barbarians* by (Changing) Partition Size	33
3.3	Nearest Neighbour for Top-rated TF-IDF Terms in *Waiting for the Barbarians*	34
3.4	Top Topics by Coherence in *Waiting for the Barbarians*	38
3.5	Top Topics by Prevalence in *Waiting for the Barbarians*	39
4.1	Developments of the Seven Types of Translation	47
4.2	Division of Work among the Four Types of Translation Technology	47
6.1	Focal Participants' Personal Profiles	79
6.2	Usefulness of Parallel Corpus	82
6.3	Search Histories on TR Corpus	82
6.4	Challenges in Using the Parallel Corpus	86
7.1	Proper Name (Single Word)	96
7.2	Reading the Proper Name as a Compound	97

8.1	Conceptual Framework with Metaverse and Social Learning	117
8.2	Groups and Scope of Study	118
8.3	Recommendations of Social Learning Methods in Metaverse Environment	124

More about the Editor

Mak Kin-wah is President of Caritas Institute of Higher Education and Caritas Bianchi College of Careers, leading the mission towards establishing St. Francis University in Hong Kong. The Institute strongly supports techno-humanities teaching and hosts a research centre dedicated to this purpose.

He has a track record of professional practice in offshore and civil engineering, as well as education and corporate management, working with government, legislators, and the media and leading major projects in Europe, Australia, Hong Kong, and Mainland China.

Dr Mak holds the degrees of PhD and MPhil from Cambridge University, an MBA from City University London, and a Bachelor of Engineering (1st Class Honours) from the University of Western Australia.

He won the Director of the Year Awards of the Institute of Directors in two consecutive years. Active in community services, he has been awarded the Bronze Bauhinia Star and is a Justice of Peace.

Contributors

Chan Sin-wai Professor-cum-Dean of the Ip Ying To Lee Yu Yee School of Humanities and Languages, Caritas Institute of Higher Education, was formerly Professor in the School of Humanities and Social Science, The Chinese University of Hong Kong, Shenzhen, and Professor and Chairman of the Department of Translation of The Chinese University of Hong Kong.

Grant **Hamilton** is Associate Professor of English Literature at The Chinese University of Hong Kong. He teaches and writes in the areas of twentieth-century world literatures in English, Anglophone African literature, computational literary studies, and literary theory. His recent book is *The London Object: Writing London at the End of Capitalism* (Routledge, 2021).

Lam Shu Yan received his BSc and MPhil in mathematical science from the Department of Mathematics at Hong Kong Baptist University, and his PhD from the City University of Hong Kong. He is currently Associate Professor at the Department of Mathematics, Statistics, and Insurance, The Hang Seng University of Hong Kong.

Lam Yee Man is currently Assistant Professor in the Department of English Language and Literature, Hong Kong Shue Yan University. She received her PhD from The Chinese University of Hong Kong. Her research interests are environmental ethics, language, and culture.

Lau Hok-Yin received his MA in Linguistics from the University of Hong Kong, specializing in sociolinguistic analyses of lyrics by use of qualitative and quantitative linguistic approaches. He is currently Lecturer at the Ip Ying To Lee Yu Yee School of Humanities and Languages, Caritas Institute of Higher Education.

Liu Jianwen is currently Assistant Professor in the Department of English Language and Literature, Hong Kong Shue Yan University. She received her PhD in Gender Studies/Translation Studies from The Chinese University of Hong Kong. Her research interests include gender-based translation studies, translation and ecology, and gender studies.

List of Contributors xiii

Liu Kanglong is currently Assistant Professor in the Department of Chinese and Bilingual Studies, The Hong Kong Polytechnic University. He received his BA from South China Normal University, his MA from Guangdong University of Foreign Studies, and a PhD from The Chinese University of Hong Kong. His research interests are literary translation and translation theory.

Thomas **Luk** Yun-tong is currently Head and Professor of the Department of English and Acting Dean of the Faculty of Arts and Social Science at Chu Hai College of Higher Learning. He received his BA in English from The Chinese University of Hong Kong, MA from York University, Canada, and PhD in Comparative Literature from the University of Michigan.

Kelly **Mak** is a Research Assistant at Chilin (HK) Ltd. She holds a BA in Linguistics and Language Application from the City University of Hong Kong, and her main research interest is Natural Language Processing.

Mak Kin-wah is President of Caritas Institute of Higher Education, which strongly supports techno-humanities teaching and hosts a research centre dedicated to this purpose. He holds the degrees of PhD and MPhil from Cambridge University, an MBA from City University London, and a Bachelor of Engineering from the University of Western Australia.

Kenny **Mok** is a Research Associate at Chilin (HK) Limited. He holds a BA in Linguistics from the University of Hong Kong, an MPhil in Linguistics from the City University of Hong Kong, and an MA in Speech Therapy from the Hong Kong Polytechnic University. His research interests include first language acquisition, language pathology, and sociolinguistics.

Ng Kwan-kwan obtained her PhD from Sun Yat-Sen University in 2018. She is currently an Assistant Professor in the Department of Chinese Language and Literature, Hong Kong Shue Yan University. She is also Director of Division of the Chinese Language Teaching and Learning, and Director of Language Centre (Chinese Section).

Shen Jiandong the Director of Soqqle Hong Kong Limited, a social mobile learning technology company that aims to encourage more social mobile learning in educational institutions through the use of mobile technology. He has more than a decade of working experience in the banking sector and felt the need for youths to learn in a more collaborative and collective way.

Su Yanfang is currently a PhD student in the Department of Chinese and Bilingual Studies, The Hong Kong Polytechnic University. Her research interests include computer-assisted language learning and corpus linguistics.

Benjamin K. **Tsou** has worked at MIT, University of California, Berkeley, University of Hong Kong, and University of California, San Diego, and founded the Research Centres on Language Information Sciences at the City University of Hong Kong and the Education University of Hong

Kong. He is the Founding President of the Asian Federation of Natural Language Processing and an Academician of the Académie Royale des Sciences d'Outre-Mer (Belgium).

Wong Pui Yun Assistant Professor in the Science Unit at Lingnan University and Centre Fellow of the Institute of Policy Studies, is a certified GIS Professional. She serves on professional associations such as the Council member and as vice chair of the Spatial Data Infrastructure Committee of Smart City Consortium.

Wong Wai Chung is Assistant Professor of the Department of Economics at Lingnan University. He received his MSc from The Chinese University of Hong Kong and PhD from Lingnan University. His current research focuses are on blended learning, housing economics, and Hong Kong economy.

Yeung Wing Lok completed his PhD in Computer Science in the United Kingdom. He has held various academic positions in universities in the United Kingdom and Hong Kong. His publications include research articles in *Science of Computer Programming*, *Formal Methods in System Design and Information*, and *Software Technology*.

Ying Koon Kau received his PhD in philosophy from the Department of Philosophy, The Chinese University of Hong Kong. He has taught in a number of tertiary institutions in Hong Kong, lecturing on philosophy and general education subjects, for nearly two decades.

Introduction

Mak Kin-wah

This book explores the relationships between technology and the humanities through case studies or specific contexts, covering a number of areas in the arts, such as language, theatre, literature, translation, philosophy, music, home designations, learning environment, and artificial intelligence. The chapters presented in this book illustrate to a great extent that the application of technology to the humanities has generated different but invariably positive results to the relevant fields.

When applied to language studies, technology brings about new horizons in techno-humanities. This is evidenced with a case study of the development of language and society by means of Big Data. In his chapter "Towards the Identification and Tracking of Salient Traits and Their Developments in Chinese Society via Big Data", Professor Benjamin Tsou Ka-yin of the City University of Hong Kong and Hong Kong University of Science and Technology, and Kelly Mak and Kenny Mok of Chilin (HK) Ltd. show with examples that the 7-billion-character Chinese corpus LIVAC database has provided information on the lexical developments associated with cultural artefacts.

When applied to theatrical arts, technology changes the performance of plays in terms of plots, themes, and settings of works. Professor Thomas Luk Yun-tong of Chu Hai College of Higher Education in Hong Kong, in his chapter "Techno-humanities: Some Trends of the Portrayal of Science in Art on the Hong Kong Stage", looks at the ways technology integrates into the theatres in Hong Kong and the changes that have resulted. A major observation made by the author is that the incorporation of technology into the theatre is closely related to the political, cultural, and economic conditions of Hong Kong, which helps the city to maintain its cultural identity.

When applied to literature, technology injects computational elements to the way works are analysed, understood, and even visualised. This is demonstrated by the chapter "Densities and Fugitive Vectors", written by Professor Grant Hamilton of the Department of English of The Chinese University of Hong Kong. He applies the proposition of Gilles Deleuze and Felix Guattari that books are an assemblage of "lines" of various kinds to analyse J. M. Coetzee's

DOI: 10.4324/9781003376491-1

experimental short novel *Foe* by treating "lines" as "vectors", thus allowing for the first time for people to understand and visualise the book-as-assemblage.

When applied to translation, technology has revolutionised the way translation is done or produced. Translation has been associated with computing from its invention in the 1940s. After a period of seven decades, translation technology has made remarkable advances in the four areas of machine translation, computer-aided translation, localisation, and speech translation. Professor Chan Sin-wai of Caritas Institute of Higher Education looks at the latest development of translation technology in his chapter "Revisiting the Future of Translation Technology", covering the period from 2013 to 2023.

When applied to philosophy, technology helps to re-examine some of the issues that have long been taken for granted. Dr Ying Koon Kau of the Ip Ying To Lee Yu Yee School of Humanities and Languages of Caritas Institute of Higher Education, in his chapter "The Idea of Techno-Philosophy and Philosophy-aided Technology, with Social Networking as an Example", raises three practical issues relating to techno-philosophy: first, the definition of "truth" in the technological world and its ontological, epistemological, and ethical implications; second, the ability of a machine to make ethical judgments; and third, the intrusion of technology to privacy.

When applied to education, technology leads to new ways of training. It is proposed by Dr Liu Jianwen of the Department of English Language and Literature of Hong Kong Shue Yan University and Dr Liu Kanglong and Ms Su Yanfang of the Department of Chinese and Bilingual Studies, The Hong Kong Polytechnic University, in their chapter "Corpus-assisted Translation Learning: Attitudes and Perceptions of Novice Translation Students", that novice translation students should be taught to use parallel corpus in translation, and suggestions have been made for future parallel corpus design and corpus-assisted translation teaching.

When applied to the naming of housing in Hong Kong, technology helps to promote the image of an "ideal home". In their chapter, entitled "What Is an 'Ideal' Home? A Multimodal Discourse Analysis of the Housing Names and TV Advertisements in Hong Kong", Dr Lam Yee Man of the Department of English Language and Literature of Hong Kong Shue Yan University, Dr Lam Shu Yan of the Hang Seng University of Hong Kong, and Dr Ng Kwan-kwan of the Department of Chinese Language and Literature, Hong Kong Shue Yan University, used Python to analyse the patterns of housing names and their advertisements from the 1980s to 2020s and found that housing naming has obliquely created an unequal social hierarchy between different classes of people in Hong Kong.

When applied to learning, technology offers a new online method to acquire knowledge. Dr Wong Pui Yun of the Science Unit and Dr Wong Wai Chung of the Department of Economics of Lingnan University, and Mr Shen Jiandong of Soqqle co-authored the chapter "A Conceptual Framework for Integrated Immersive Learning with Metaverse and Student-generated Media" to review the use of immersive technologies in a multi-disciplinary university.

When applied to theoretical concepts, technology generates new thoughts on its role. In his chapter "Problems of Exacerbation to Dasein in the Modern Technological World by Use of the Early Heidegger's Theories: Readiness-to-hand and Presence-at-hand", Mr Jeff Lau Hok-yin of Caritas Institute of Higher Education argues that, from Heidegger's theories of readiness-to-hand and presence-at-hand, it could be seen that there are three types of problems relating to technology: (1) technical problems, (2) non-technical problems, and (3) problems to Dasein's being-there brought by technology.

When applied to artificial intelligence, technology sheds light on the emotional aspects of computational applications. Dr Yeung Wing Lok of the Techno-Humanities Research Centre of Caritas Institute of Higher Education discusses the ethics of computing through empathy in AI Chatbots in his chapter "Ethically Speaking: Opportunities and Risks of AI Chatbots Showing Empathy to Customers during Service Encounters". His conclusion is that empathy could be good if it is not perceived as fabricated and that there are risks of artificial empathy, which includes anthropomorphism.

These chapters are not the end of our discussion on the advances of techno-humanities; rather, they serve as a new starting point for us to explore more deeply the role of technology in the humanities.

1 Towards the Identification and Tracking of Salient Traits and Their Developments in Chinese Society via Big Data

Benjamin K. Tsou, Kelly Mak, and Kenny Mok

Rapid developments with the cultivation of Big Data and their applications are not only hallmarks of recent technological advancements but also open up new horizons for digital humanities. Sociolinguists have made use of Big Data to fruitfully examine our everyday language usage to explore the underlying social structure and cultural values. In this chapter, we take a new approach by focusing on some areas of concern to social scientists and humanists and by making use of the Big Database LIVAC to explore relevant latitudinal and longitudinal developments in the Pan-Chinese communities. LIVAC is a 7-billion-character Chinese corpus which has continuously and synchronously drawn on media materials from six Chinese speech communities since 1995. It is noteworthy that this uncommon corpus has made possible the discovery of new trends and consequential developments such as socio-economic satisfaction, self-identity and ethnicity, cultural exchange, and gender equality issues within the Pan-Chinese communities on the basis of the mutual relationship between language and the larger societal and cultural contexts. Furthermore, the period of concern is one of colossal social transformation and economic advancement.[1]

The identification and tracking of salient traits and developments in society are of primary concern not only to the social scientists and humanists but also to the other informed members of society. Popular methodologies involve two major components: (1) data source by means of direct collection, through conducting surveys among respondents who provide the direct source of information, and (2) the subsequent analysis of the data relevant to the targeted salient traits. It is also recognized by many that language is intimately associated with the complex and extensive social interactions unique among human beings. As such, they reflect the underlying intention, motivation, and reaction of the speakers through their verbal and other behaviours and thereby reveal some of the conscious and unconscious aspects of their cultural and social values and ideologies. A range of useful analysis can be undertaken, which could possibly vary from obtaining simple quantitative distributions to more complex extrapolation of salient traits. For example, speakers of Chinese, Japanese, and Korean would use euphemistic expressions or avoid using homonyms of the word for death, "si", in naming people and objects, because

DOI: 10.4324/9781003376491-2

The Identification and Tracking of Salient Traits 5

of the social taboo concerning death. For example, the Chinese verb dying "si" is usually replaced by "qu shi" (*to depart from the world*), and the noun referring to a dead person "si ren" is usually replaced by "xian ren" (*the person who preceded us in this world*). In some societies, the impact may be extended beyond a personal level of word choice in naming by speakers. For example, a building may skip the numbering of the 4th, 14th, and higher floors with the number "four" because "si" is a homonym of the word for death. In Japanese enumeration, the use of the Sino-Japanese word "si" is replaced by the native Japanese word "yon" due to the taboo. These examples are good illustrations of how language has a strong association with its users and offers revealing glimpses of salient social and cultural traits among individual groups of speakers and the society as a collective whole. Further examples could provide more profound revelations about the speakers and their societies, such as their beliefs, orientations, and sentiments in respect of abstract or concrete subjects, as will be discussed subsequently in this chapter.

Two major types of data source can be relevant to the investigation of specific targets. The first is firsthand data or data directly obtained from the subject. The second is indirect data which could not be directly solicited from the subject but which would have to result from data mining to obtain new derivative data that have been indirectly collected and objectively analysed. An example of directly solicited data is the data obtained through questionnaires amongst members of a society on their sentiments towards the quality of life or the environments. The investigation could, for example, ask for the respondents' positive or negative feelings about the life they lead and the living conditions they have. This could be done within a single community, and the results may be also compared with those from other communities. Despite the usefulness of this kind of data, they may have limitations such as the nature and background of the selected respondents, and the extent of true representation. There are further restrictions in terms of timeliness because it is impossible to go back in time to find out how subjects might have retroactively responded to the same questionnaire. In short, longitudinal study is not possible unless it is well planned. What is readily possible would be contemporaneous latitudinal comparisons where qualified respondents may be more readily available.

For the purpose of pursuing longitudinal comparisons, we could make use of accumulated information over time, typically written records, and analyse representative media materials over the relevant period, when the collective sentiments or feelings in society have been filtered and readily identified. They are significant sociocultural markers often found encoded in the language. In Chinese, for example, adjectives and adverbs are the language categories which can overtly represent the speaker's sentiment. However, at a deeper level, Chinese idioms, or quadrisyllabic idiomatic expressions (QIEs), can also reveal the speaker's underlying sentiments towards certain objects or topics. Take the example of 今非昔比,[2] "not like what it used to be before".

It can be seen from Table 1.1 that this popular QIE not only has the literal meaning of comparison between the current and past conditions but also

Table 1.1 Positive and Negative Connotations of QIE 今非昔比 "not like what it used to be before" Based on LIVAC

Positive

Beijing
1. 沙鋼已<u>今非昔比</u>，成為江蘇冶金行業第一家省級集團。

    "Shagang is <u>not like what it used to be before</u>; it has become the first provincial group in Jiangsu's metallurgical industry."

2. 先鋒村用發展生產和節省的吃喝費辦了公益事業，使先鋒村發生了<u>今非昔比</u>的變化，一幢幢磚瓦房替代了過去的茅草屋。

    "With the money saved from production and from entertainment to run public welfare, Pioneer Village underwent changes with brick houses replacing the old, thatched cottages, making it <u>not like what it used to be before</u>."

Hong Kong
1. 隨著十多年急速的經濟增長，中國的國力早已<u>今非昔比</u>，現已儼然成為亞洲的龍頭。

    "With more than a decade of rapid economic growth, China's national strength is <u>not like what it used to be before</u>, and she has now become Asia's leader."

2. 中國對土耳其十分看重，認為該隊現時有多位球員外流歐洲 (已歸隊)，實力已是<u>今非昔比</u>，將是中國隊的一大威脅。

    "China is very keen on Turkey, believing that with the return of many of its exited players from Europe (to the national team), it is truly <u>not like what they used to be before</u> and will be a major threat to China."

Shanghai
1. 埃弗頓隊主教練莫耶斯也不示弱："埃弗頓隊早就<u>今非昔比</u>了，誰輕視他，誰就是自找麻煩！"

    "Everton coach Moyes is also not showing weakness: 'Everton is already <u>not like what it used to be before</u>. Whoever despises Everton is asking for trouble!'"

2. 看到裝備先進的精紡車間，打趣地說："你們這是<u>今非昔比</u>，鳥槍換炮啦。"

    "When he saw the state-of-the-art precision spinning workshop, he jokingly said, 'You're <u>not like you used to be before</u>; The shot guns have become cannons.'"

Taiwan
1. 其二，是北韓<u>今非昔比</u>，已經擁有核子武器與製造飛彈的技術。

    "The second is that North Korea with its nuclear weapons and missile technology, it is <u>not like what it used to be</u>."

2. 當時盧彥勳止步16強，盧彥勳說，今非昔比，有信心在今年討回顏面。

    "At that time, Lu Yen-hsun stopped in the eighth-finals. He said that he is <u>not like what he used to be before</u> and was confident that he would achieve a big turnaround this year. "

Table 1.1 (Continued)

Negative

Beijing
1. 週邊環境<u>今非昔比</u>，儘管集團一再促銷，出租率仍然不佳。

    "The surrounding environment is <u>not like it used to be before</u>; despite repeated promotions, the occupancy rate remained poor."

2. 客隊<u>主力流失</u>，實力今非昔比。
    "The strength of the visiting team is <u>not like it used to be before</u> with the loss of main players."

Hong Kong
1. 但黃柏高承認，樂壇<u>今非昔比</u>，現在歌手出碟、拍電影是宣傳，拍廣告、演唱會才是賺錢。

    "But Paco Wong admits that the music industry is <u>not like it used to be before</u>. Nowadays, singers' albums and movies are for promotion, while advertisements and concerts are for making money."

2. 她以為可自此佔盡地利優勢，可是<u>今非昔比</u>，被無牌小販搶去不少生意。

    "She thought she could take advantage of the location but now is <u>not like what she to be before.</u> She has lost a lot of business to unlicensed hawkers."

Shanghai
1. 賽後，擔任比賽監督的中國隊教練施之皓說："今天的上海隊已<u>今非昔比</u>，今年恐怕要為保級而戰了。"

    "After the match, China coach Shi Zhihao, who supervised the match, said, "Shanghai today is <u>not like what it used to be before</u>, and I am afraid they will have to fight for relegation this year.""

2. 法國能源與運輸集團——阿爾斯通公司真是<u>今非昔比</u>：連續兩年虧損；債務累計高達近50億歐元。

    "The French energy and transport group, Alstom, is really <u>not like what it used to be before</u>: two consecutive years of losses, a cumulative debt of nearly 5 billion Euros."

Taiwan
1. 而曾經神氣活現的國王隊（7-8），退居倒數第2，只比<u>今非昔比</u>的湖人（5-8）略勝一籌。

    "The Kings (7–8), who were once exalted, relegated to the bottom two, only slightly better than the Lakers (5–8), which <u>were not like what they used to be before.</u>"

2. 相比「沒什麼娛樂、看電影就是最大的娛樂」的美好年代，簡直令人有<u>今非昔比</u>之歎。

    "Compared to the good old days when 'nothing was more entertaining than watching movies', we can only heave a sigh that is <u>not like what it used to be before.</u>"

8 *Benjamin K. Tsou, Kelly Mak, and Kenny Mok*

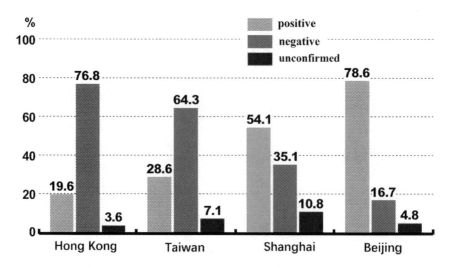

Figure 1.1 Sentiment Polarity of the QIE 今非昔比 "Not like what it used to be before" across Four Pan-Chinese Communities in 2005–2013.

reflects deeper sentiments by the speakers as the QIE is used. It can have both positive and negative connotations in all Chinese speech communities. However, Figure 1.1 shows that there are significant differences in the overall connotations amongst the four major Chinese communities within an eight-year time span.

Figure 1.1 shows a spectrum of positive and negative sentiments in each community, but an overall contrast is very much evident among leading cities in mainland China, such as Beijing and Shanghai, which use it mostly with positive connotation, on the one hand. Meanwhile, on the other hand, those outside, such as in Hong Kong and Taipei, use it mostly with a negative connotation. The follow-up focus group discussions with participants from the four cities have yielded reports of differential sentiments about lives in general in the two regions, especially about changing income variations and living conditions in the period concerned.

It would be much harder to compare similar findings over time with direct questionnaires as there would be practical constraints, such as the selection and retroactive availability of comparable respondents in the different communities with respect to age, income, and gender, for example.

A related example may be an investigation of self-identity and ethnicity in Hong Kong in conjunction with the pre- and post-1997 changeover on sovereignty. An investigator could hold a survey on ethnicity amongst a single group of respondents over a ten-year period straddling 1997 and obtain useful results by asking, for example, how the respondents feel towards their motherland in

Table 1.2 Reference to Chinese National Leaders in the HK Media (1995–2002)

	總理 premier		國家主席 president	
Year	With "China" (%)	Without "China" (%)	With "China" (%)	Without "China" (%)
95	58.33	41.67	93.75	6.25
96	67.80	32.20	85.90	14.10
97	48.08	51.92	57.14	42.86
98	26.13	73.87	40.28	59.72
99	31.15	68.85	41.98	58.02
00	17.46	82.54	22.62	77.38
01	4.05	95.95	22.37	77.63
02	8.33	91.67	19.05	80.95

Note: For more details, see Tsou and Kwong (2015a).

terms of a range of questions related to their self-identification as a Chinese. The survey sampling representativeness, the understanding of the survey questions by the participants, as well as the truthfulness of their responses, are always subject to question in the direct method of data collection, especially for investigations of this kind of topic. However, it is worthwhile to look into individual and societal views about self-identity issues by means of the language usage of the speakers and the media in the society, which indirectly also reflect their attitudes. As shown in the Table 1.2, it can be seen that the use of "China" as an attribute before the official title of the country's Premier or President in the news media has changed incrementally over time from pre- to post-1997.

It is noteworthy that this kind of survey and comparison cannot be conducted retroactively. Thus, if, as an afterthought on the part of an investigation, there is interest in comparing the Hong Kong residents' identification with China as compared with Britain over time, the investigator could be disappointed in not being able to conduct a survey retroactively after 1997. The methodology adopted in this chapter would provide useful information as shown in Table 1.3, where it will be seen that the references to political leaders from Britain have always had the attribute "British", thereby showing consistency in a more distant ethnicity with reference to Britain.

Similarly, the availability of a synchronous news media database would allow the rigorous and fruitful retroactive comparison of ethnicity between Hong Kong and other communities such as Macau and Taiwan within the designated period. It should be further noted that the approach taken in this chapter is primarily suitable for global comparison because the media could provide a bird's-eye view on the society. It would not provide a fine-grained analysis with variables such as age, gender, or social background of subgroups in the society.

Table 1.3 Attributes of Proper Names in Hong Kong (Pre- and Post- 1997)

中國 China	英國 Britain (Constant)
1. (中國)國家主席 江澤民/ 胡錦濤 (China) President Jiang Zemin/ Hu Jintao	
2. 李鵬總理/ 朱鎔基總理/ 溫家寶總理 Premier Li Peng/ Zhu Rongji/ Wen Jiabao	英國首相 British's Prime Minister
3. 李總理/ 朱總理/ 溫總理 Premier Li/ Zhu/ Wen	梅傑/ 布雷爾/ 布朗/ 卡梅倫 Major/ Blair/ Brown/ Cameron
4. 李總/ 朱總/ 溫總	

New knowledge can be further gained through extrapolation and analysis after data mining, which cannot be obtained from the direct data collection method. This kind of knowledge discovery is of fundamental importance in the use of Big Data in digital humanities. For example, we may be interested in traditional macro questions such as the mutual influence in many spheres between the Chinese mainland on the one hand and Hong Kong and Taiwan on the other hand in the run-up to 1997 and thereafter. If we use the direct data method, it will be an excessively mammoth and complicated project, requiring tight control over time and over the cultivation of a new big database that has to provide longitudinal information. If the interest is focused on trade and economics, the Big Data approach would be based on related familiar economic indicators, such as CPI, GDP, and others. The results of such an analysis of the Pan-Chinese communities would show the spectacular rise of the economy in mainland China and its mutual influence relative to Hong Kong and Taiwan.

Compared to economic developments, there can be no less significant cases concerned with developments in cultural and their mutual influence among the Chinese communities. We could make use of neologisms as a class of cultural markers to observe the rise and transfer of new cultural artefacts in the different Chinese communities.[3] The results of such a study for Hong Kong and mainland China are given in Figure 1.2.[4]

The data for each year are obtained by examining new words in both Hong Kong and mainland China's Shanghai and Beijing, where many had been found in one of the communities previously and so could contribute to an indication of the direction of influence for comparison. These results show that between 1995 and 2015, there have been interesting and contrastive developments in the northbound and southbound traffic of the new terms and the new ideas they represent. Some notable examples are provided in Table 1.4.

These examples show that southbound terms are mostly related to science and technology, as well as communications, while the northbound terms

The Identification and Tracking of Salient Traits 11

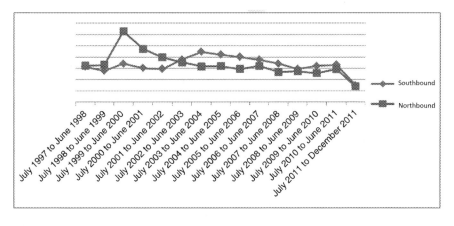

Figure 1.2 Overall Northbound and Southbound Lexical Transfers between Mainland China and Hong Kong (1997–2011).

Table 1.4 Lexical Transfer Northbound and Southbound (1997–2011)

內地南下詞語 *Lexical Transfer Southbound*	香港北上詞語 *Lexical Transfer Northbound*
1 微博Weibo	訴求Appeal
A social media platform in China	A serious request to deal with problems
例句：郭子威的微博有超過1300名「粉絲」，其長實地產投資董事身份通過新浪網認證。	例句：主席田北辰說，這已是九鐵可以提供的最大優惠，並已回應市民訴求，紓解民困。
"Guo Ziwei has over 1,300 "fans" (followers) on his Weibo account, and his identity as a director of Cheung Kong Property Investment is verified through Sina.com."	"Chairman Michael Tian said that this was the biggest concession that the KCRC could offer and has responded to the public's appeal to relieve people's hardships"
2 維權Rights Protection	問責Accountability
to defend (legal) rights	the expectation of the proper discharge of responsibilities
例句：據了解，廈門市電業局引入公證依法維權的舉措，已引起了國內其他供電、公證部門的濃厚興趣。	例句：該署強調，十分關注有關指控，並必須向公眾問責。
"It is understood that the introduction of notarization and rights protection by Xiamen Electric Power Bureau has aroused strong interest from other domestic power supply and notarization departments."	"The Department stressed that it was very concerned about the allegations and the need to be accountable to the public."

(*Continued*)

Table 1.4 (Continued)

內地南下詞語 *Lexical Transfer Southbound*	香港北上詞語 *Lexical Transfer Northbound*
3　黃金周 Golden Week a week-long national holiday 例句：黃金周期間，全國旅遊秩序普遍健康有序，旅遊者的出遊滿意率超過90％。 "During the Golden Week, tourism across the country was generally healthy and orderly, with the satisfaction rate of travelers exceeding 90%."	解讀 (Legal) Interpretation consequential opinion on an issue of importance 例句：李耀培表示，每宗意外的事發原因、求情理由以至法官對案件的解讀均有不同，難以評論本案判刑是否合理。 "Ringo Lee said that it was difficult to comment on the reasonableness of the sentence in this case as the cause of each accident, the reason for the plea and the judge's interpretation of the case differed."
4　減排 Emission Reduction to reduce the amount of pollutants 例句：該建議允許發達國家用更多的森林植被等抵消溫室氣體減排指標。 "The proposal allows developed countries to offset greenhouse gas emission reduction targets with more forest cover, for example."	減持 Reduce Holding selling out stocks 例句：對於早前傳出摩根士丹利沽空期指，其發言人稱公司未有減持港股，仍然維持年底恆指可達萬九點的預測。 "Regarding the rumors of Morgan Stanley short-selling the future index earlier, its spokesman said that the company has not reduced its holdings of Hong Kong stocks and still maintains the forecast that the Hang Seng Index will reach 19,000 by the end of the year."
5　自由行 Individual Visit Scheme a visit without supervision of guide 例句：旅遊發展局主席周梁淑怡表示，歡迎內地開放更多省市的遊客自由行來港。 "The Chairman of the Tourism Board, Mrs Selina Chow, said she welcomed the opening up of more provinces and cities from the Mainland to visit Hong Kong under the Individual Visit Scheme."	雙贏 Win-win of or denoting a situation in which each party benefits in some way 例句：馬會主席夏佳理與澳門馬會主席何鴻燊經已達成共識，並擬定雙贏合作方案。 "Jockey Club Chairman Ronald Arculli and Macau Jockey Club Chairman Stanley Ho have reached a consensus on a win–win cooperation proposal."

Table 1.4 (Continued)

內地南下詞語 *Lexical Transfer Southbound*	香港北上詞語 *Lexical Transfer Northbound*
6　提速 To accelerate to speed up with consequence 例句：停滯不前的北京市危舊房改造工作走出了"死胡同"，進入了全面提速階段。 "The stagnant renovation of dilapidated houses in Beijing has come out of a 'dead end' and entered a stage of overall acceleration."	追捧 To chase after hotly pursue (stocks, etc.) 例句：除了泛海酒店，部份酒店股確有買家追捧，大酒店及富豪酒店分別上揚百分之三點四及百分之四點八。 "Apart from Pan Ocean Hotel, some hotel stocks did see a surge in buyers, with Grand Hotel and Regal Hotel rising by 3.4% and 4.8% respectively."
7　經濟體 Economic Community an entity or organization involved with production, distribution and trade, as well as consumption of goods and services 例句：由於北韓實行高度計劃經濟，且是世界上最孤立的經濟體之一，政府不對外公佈其真實的經濟數據。 "As North Korea is a highly planned economy and is one of the most isolated economies in the world, the government does not publish its real economic data to the public."	爆冷 Unexpected development a situation in which a person or team beats the person or team that was expected to win 例句：沙田馬場昨晚舉行的七場夜馬賽事雖然爆冷頻頻，但終能順利進行，並無意外發生。 "The seven night races held at the Shatin Racecourse last night went off without a hitch, despite a number of upset results."
8　國足 National Football Team a team that represents a country, rather than a particular club or region, in an international sport 例句：在昆明集訓期間，國足並沒有以韓國隊為重點進行針對性訓練。 "During the training camp in Kunming, the national football team did not focus on the South Korean team for targeted training."	操控 Manipulation controlling someone or something to own advantage, often unfairly or dishonestly 例句：同時，為免出現造假波，馬會在首階段會避免賭一些容易受操控的項目。 "At the same time, the Club will avoid betting on easily manipulated events in the first phase to avoid any fraud."
9　博客 Blog a regular record that is put on the Internet for public consumption	惡搞 Fooling Around a silly composition or effort, often poking fun at an original work

(*Continued*)

Table 1.4 (Continued)

內地南下詞語	香港北上詞語
Lexical Transfer Southbound	*Lexical Transfer Northbound*
例句：隨著互聯網技術的快速發展，博客、論壇、搜索引擎等的使用越來越廣泛，社會影響力越來越大。 "With the rapid development of Internet technology, the use of blogs, forums, and search engines is becoming more and more widespread and socially influential."	例句：該片開拍至今爭議不絕，網民紛紛藉機惡搞並喻諷當今時局。 "The film has been controversial since its inception, with netizens taking the opportunity to mock the current situation."
10 高新 High-tech technology that is at the cutting edge	智庫 Think Tank a research institute concerned with such topics as social policy, political strategy, economics, military, technology, and culture
例句：他說，中國的市場很大，發展高新技術產業的前景非常廣闊，希望大家繼續努力，取得更大的成績。 "He said that China has a huge market, and the prospects for developing high-tech industries are very broad. He hoped that everyone will continue to work hard and achieve greater achievements."	例句：日前，有本地智庫組織請來史丹福大學的教授，發表關於香港教學語言的演講。 "Recently, a local think tank invited a professor from Stanford University to give a talk on the language of instruction in Hong Kong."

Table 1.5 Percentage Distribution of Terms with the Two Gender Headwords in Beijing, Hong Kong and Taipei (1996–2000 and 2011–2015)

%	Beijing		Hong Kong		Taipei	
	96–00	*11–15*	*96–00*	*11–15*	*96–00*	*11–15*
Male (男) Headword	0	39.2	0	59.1	9.1	52
Female (女) Headword	100	60.8	100	40.9	90.9	48

are more in the field of economics and social well-being. The social and cultural imbalance observed provides a good glimpse of the directions of future developments in the two regions. A similar kind of imbalance found in the economic sphere will certainly draw much interest and the concern of politicians and policymakers. It may be argued that these findings, in exploring the social and cultural sphere and in the methodology deployed, might deserve no less interest among the social scientists and policymakers.

A further example may be gender (in)equality and variations over time. To explore issues on gender (in)equality, the use of the traditional deductive approach, as well as direct solicited method such as questionnaires and surveys, has been adopted. For example, governmental organizations, such as the Equal Opportunity Commission, frequently adopts this direct method approach to collecting respondents' subjective self-reported perception in real-life contexts for investigations.

On the other hand, indirect methods could be profitably used, such as analysing the gender pay gap. Based on reports in the film industry in Hollywood, there are reports that a well-known actress such as Jennifer Lawrence earned $5 million less than her co-star Leonardo DiCarpio in the same movie (Smith 2022). Other collaborating evidence has shown that, generally, actresses are paid less than actors[5] (Izquierdo Sanchez and Paniagua, 2017), thereby confirming the existence of gender discrimination for cultural icons on the basis of gender.

Language use in discourse as a tool to study gender bias in English was seriously taken up by Lakoff (1975) and Spender (1980). For instance, the words "master" and "mistress" were originally neutral, masculine, and feminine honorific terms to describe someone with a professional status in the past. However, only the feminine term has evolved to acquire an extended sexual meaning of "paramour". The additional negative connotational meaning is not found for the male counterpart "master".

(Lakoff [1975])

(1) (a) He is a master of the intricacies of academic politics.
    (b) *She is a mistress ...

(2) (a) *Harry declined to be my master and so returned to his wife.
    (b) Rhonda declined to be my mistress, and so returned to her husband.

A case of asymmetry can also be found from a qualitative analysis of Chinese terms with the headwords 男 "male" and 女 "female" (See Table 1.6). By means of a computational analysis of the Pan-Chinese media corpus LIVAC, we have observed that, from 1996 to 2015, 男 "male" was rarely used as an attribute in Beijing, Hong Kong, and Taipei (in 1996–2000), as shown in

Table 1.6 Connotations of Top Frequency Terms with the Two Gender Headwords

	Male	*Female*
+	31	6
−	23	48
N	12	12

Table 1.7 Examples of Top Frequency Terms with the Two Gender Headwords in LIVAC

Male (男) Headword					
1	型男	Handsome man	11	潮男	Trendy man
2	宅男	Otaku	12	港男	Hong Kong man
3	猛男	Strong man	13	賤男	Mean man
4	少男	Young man	14	惡男	Wicked man
5	美男	Beautiful man	15	熟男	Mature man
6	俊男	Handsome man	16	嫩男	Young man
7	壯男	Strong man	17	醜男	Ugly man
8	肌肉男	Muscular man	18	處男	Virgin man
9	裸男	Naked man	19	內地男	Mainland man
10	剩男	Leftover man	20	舞男	Gigolo
Female (女) Headword					
1	妓女	Prostitute	11	智障女	Retarded girl
2	美女	Beauty	12	宅女	Homebody girl
3	靚女	Pretty girl	13	烈女	Virtuous woman
4	港女	Hong Kong girl	14	按摩女	Masseuse
5	修女	Nun	15	歌女	Songstress
6	處女	Virgin	16	舞女	Dancehall girl
7	剩女	Leftover woman	17	聖女	Holy woman
8	裸女	Naked woman	18	潮女	Trendy woman
9	富家女	Rich woman	19	豪放女	Wild girl
10	醜女	Ugly girl	20	拜金女	Gold digger

Table 1.5. However, the male gender group, in the later period (in 2011–2015) gained rising attention.

On the other hand, quantitative comparison shows that for an equal number of top frequency terms, there are more positive than negative terms with the headword "male". At the same time, the opposite results are found for the headword "female", where the contrast is even greater. Table 1.7 shows some of the top frequency terms with the two headwords. It is noteworthy that most attributes of the male terms are positive comments on their appearances (e.g. handsome, strong), while more negative terms are related to the sex industry (e.g. prostitute, massage girl) for those with the female headword. Such bipolar connotations on attributes demonstrate Chinese societies' values for female virtue and higher moral standards for females than for males.

The diverse range of findings on same salient traits in Chinese society reported here have been made possible by means of latitudinal and longitudinal comparisons involving the Big Database LIVAC, which deals with the use of the Chinese language in the Pan-Chinese contexts.[6] The project was launched in 1995 and since then has cultivated more than 700 million characters of news media texts in the Pan-Chinese communities, including Beijing, Hong Kong, Macau, Shanghai, Singapore, Taiwan, and others. Partly because its rigorous curation included assisted verification, LIVAC not only has been able to provide fruitful and reliable exploration of the linguistic variations across time and space but has also enabled a fuller account to be made of the wider underlying societal and culture cross-currents.

Notes

1 Research reported in this chapter has benefited from the support of many sources, including the initial grant from the CCC Foundation in Taipei and others from the Research Grants Council of Hong Kong (7000818, 9040782, 9040860, 7002032, 9041193, 9667013, 9440065, 844012), as well as from the Language Fund and Lord Wilson Trust of Hong Kong. Some parts of this chapter were reported in a number of related publications and presentations, and a note of appreciation must be recorded for the abundant insightful feedback received, though the authors alone must be responsible for its content.
2 See Tsou (2013) and (2016) for more details on the cultural saliency of these constructions.
3 Tsou, B.K. (2004). "Towards a Comparative Study of Diachronic and Synchronic Lexical Variation in Chinese," in *Mapping Meanings the Field of New Learning in Late Qing China*, edited by M. Lackner and N. Vittinghof, Berlin: J Brill, pp. 355–77.
4 For more details, see Tsou and Kwong (2015a).
5 In other review of the highest-earning actors for a single production, there is only one female actress, Sandra Bullock (ranked sixth) in *Gravity* (2013), among the top ten actors. The highest income record for Keanu Reeves in *The Matrix* (2003) sequels is two times Sandra's. Furthermore, if we expand our scope to the 20 highest-earning actors, there is only one other actress on the list (i.e. Cameron Diaz in *Bad Teacher* [2011]) (https://en.wikipedia.org/wiki/List_of_highest-paid_film_actors).
6 https://en.wikipedia.org/wiki/LIVAC_Synchronous_Corpus. See also Tsou et al. (2011), Tsou and Kwong (2015), and Tsou (2022a,b).

References

Izquierdo Sanchez, Sofia and Maria Navarro Paniagua (2017) *Hollywood's Wage Structure and Discrimination*, https://doi.org/10.13140/RG.2.2.29148.18562

Lakoff, Robin Tolmach (1975) *Language and Women's Place*, New York: Harper & Row.

Smith, Morgan (2022) *Jennifer Lawrence Slams Hollywood's Gender Pay Gap: 'It Doesn't Matter How Much I Do'*, CNBC, Retrieved 18 November 2022, from www.cnbc.com/2022/09/07/jennifer-lawrence-slams-hollywood-gender-pay-gap-in-vogue-interview.html

Spender, Dale (1980) *Man Made Language*, London: Routledge and Kegan Paul.

Tsou, Benjamin K. (2004) "Towards a Comparative Study of Diachronic and Synchronic Lexical Variation in Chinese", in Michael Lackner and Natascha Vittinghof (eds.) *Mapping Meanings the Field of New Learning in Late Qing China*, Berlin: J Brill, 355–77.

Tsou, Benjamin K. (2013) "Emblematic Chinese Quadrasyllabic Idiomatic Expressions (QIE's) in Asian Languages: Explorations in Linguistic Reconstructions and Cultural History, and New Proposal for UNESCO Recognized Intangible Cultural Heritage",

Keynote paper presented at The 21st Annual Conference of the International Association of Chinese Linguistics (IACL-21), Taipei.Tsou, Benjamin K.鄒嘉彥 (2016). "粵語四字格慣用語：傳承與創新". 鄒嘉彥 湯翠蘭 (2017). 《中國南方語言四音節慣用語研究》《粵語研究》增刊. 鄒嘉彥 李錦芳 (2016) 編《百越語言研究：中國南方語言四音節慣用語論文集》 (*Papers on "Quadra-syllabic Idiomatic Expressions in Languages around South China"*), 香港. pp. 139–157.

Tsou, Benjamin K. and Kwong Oi Yee (eds) (2015) *Linguistic Corpus and Corpus Linguistics in the Chinese Context (Journal of Chinese Linguistics Monograph Series No.25)*, Hong Kong: Chinese University Press.

Tsou, Benjamin K. and Kwong Oi Yee (2015a) "LIVAC as a Monitoring Corpus for Tracking Trends Beyond Linguistics" (從LIVAC追蹤語料庫探索泛華語地區語言以外的演變), *Journal of Chinese Linguistics Monograph Series*, 25 (2015/12/01), 447–71.

Tsou, Benjamin K. and Yip Ka-Fai (2020) "A Corpus-based Comparative Study of Light Verbs in Three Chinese Speech Communities", in *Proceedings of the 34th Pacific Asia Conference on Language, Information and Computation, PACLIC*, 302–11.

Tsou, Benjamin K. 鄒嘉彥, Kwong Oi Yee 鄺藹兒, Lu, Bin 路斌 and Tsoi Wingfu 蔡永富 (2011)〈漢語共時語料庫與追蹤語料庫：語料庫語言學的新方向〉(Chinese Synchronous Corpus and Monitoring Corpus: A New Direction of Corpus Linguistics),《中文信息學報:慶祝中國中文信息學會成立三十周年紀念》(*Journal of Chinese Information Processing "Special Issue to Commemorate 30th Anniversary of Chinese Information Processing Society"*), 25(6):38–45. China: Chinese Information Processing Society.

Tsou, Benjamin K. (2022a) "Some Recent Trends in Chinese Language and Society from the Prospective of Big Data: Beginning with Hong Kong" ([從大數據庫看泛華語區語言社會何去何從：從香港說起], Keynote paper presented at the 22nd International Conference on Chinese Language and Culture (ICCL -22), Singapore.

Tsou, Benjamin K. (2022b) "Some Saliant and Divergent and Convergent Linguistic Developments in Chinese – A Big Data and Trans-Millennial Approach", Keynote paper presented at The 28th Annual Conference of the International Association of Chinese Linguistics (IACL-28), Hong Kong.

2 Techno-humanities
Some Trends of the Portrayal of Science in Art on the Hong Kong Stage[1]

Thomas Luk Yun-tong

Technology and the Rise of the Playhouse

The rise of the playhouse followed, as a rule, the flourishing of dramatic literature since the very beginning of theatre history. From the very early phase of the theatre, which rested solely on the player and the space, to its gradual accretions in architecture, scenography, and modern technologies, the steady influence of the dramatic literature on the stage and the performing area's subsequent turning indoors have continued hand in hand. Today, technological influence has firmly wedged its foot in the door, as the theatre's gravitation towards technologies started inevitably with the receding of the proscenium arch in the eighteenth century into a box-like frame and with the increasing utilization of stage machinery, electric lighting, the revolving stage, videoclips, and sound recordings systems, etc., run by the computer.

Though some theatre scholars may insist that "theatre building and its concomitant technologies have seldom given birth to new forms of dramatic art that have had a life or significance beyond the immediate entertainment and gratification of its contemporary audience,"[2] the development seems to subvert this claim in contemporary theatre, which has witnessed more frequent use of multimedia and other appropriations of technologies as new forms of theatre practice. The uses range from the mere decorative backdrop to exhibit a period or cyber outlook, as in the case of *Miss Saigon* (1989), to a thematic exploration of human conditions and technology, as *In the Wire*, a play about turning mysteries of email and technology into a storyline on the stage, presented in August 2002 at the New York International Fringe Festival.[3]

Contemporary theatre since The Wooster Group has incorporated different technologies into theatre form and introduced two performance languages live and mediatized in the dramaturgical process. The use of technology and intermedia matters, especially in the form of video projection or clips, film footage, and recorded music, has caught on in contemporary Hong Kong theatre as well, not only among avant-garde or semi-professional groups such as Amity Drama Club but also among major flagship theatre companies, such as Hong Kong Repertory Company and Zuni Icosahedron.[3,4] This integration

DOI: 10.4324/9781003376491-3

of technology and art, thematically, aesthetically, and theatrically, has become a trend on the Hong Kong stage, aside from appropriating or adapting traditional Chinese opera or Western classics. The presence of intermediality in the *mise en scène* is steadily generating a new type of dramaturgical form of art, provided by stage technologies for exposition, narration, and juxtaposition as well as the dramatization of complex subject matter such as human nature and behaviour. Technologies offer the theatre the nutrition to examine who we are, to raise questions about personal identity, to discern boundaries between living things and machinery, to learn about communication and alienation. In the following, I shall explore this trend of intermediality on the Hong Kong stage concerning some local productions in the past years.

Before doing so, a brief introduction to modern/contemporary Hong Kong drama is in order. Modern/contemporary spoken drama in Hong Kong, like its counterpart in China, came from the same source, the West, at the beginning of the twentieth century, spanning a history of about a hundred years, and has been coterminous with the rise of a naturalistic theatre with Ibsen-Shaw-Chekhov tributaries in the West. The art form is distinct from classical Chinese theatre, which has a long history of non-illusory performance that has continued to the present day. Yet contemporary Hong Kong theatre is post-modernist, in a chronological and generic sense, as it bespeaks a theatre from the 1960s onward, experimental in outlook, exploratory and intellectual, often self-consciously interested in its identity status and formation. Besides appropriating or reinterpreting traditional themes and matter, as well as nourishing itself with Western dramaturgy through productions of original and translated plays, some of the works of contemporary Hong Kong theatre accentuates a post-modernist aura by ostensibly having made multimedia technology a regular factor or theme in their productions.

Technology-inspired Productions on the Hong Kong Stage

Let me first call your attention to two of the productions with historical reflection and contemporary topicality by The Amity Drama Club (致群劇社), a very time-honoured dramatic group in Hong Kong. The first production is *The Search for the Spring Willow* (1997), which puts a local repertory on stage for self-reflection on its artistic development and direction; it incorporates aspects of Cantonese opera, Western dramatic text, and cinematic/intermedial techniques to show a concern beyond the theatre and a historical galloping survey of modern Chinese history since the Late Ch'ing period (1908–11) to modern-day Hong Kong.

Dramaturgically, this production resorts to multimedia technology, as video interviews of personalities and movie clips of self-narration on the stage serve expository and narrative functions, offering multiperspective, multilayered expressions, time and space crisscrossing, and montage juxtaposition and highlighting theatre as a means of representing illusion and reality as well as contextualizing its self-conscious objective: an intermedial exploration of theatre in the age of technology.

Staged in 1997, the play is structurally divided into two parts. Part one is set against a Hong Kong local repertory, with its director as a spokesperson searching for artistic excellence. He is a Socrates expounding on the essence of the theatre and art to his drama pupils, as the Russian director Stanislavsky does in his book *Building a Character*. The second part dramatizes the return of students from Japan at the turn of the last century and their establishment of the Spring and Willow Society, in search of a practical form of theatre for the betterment of China and its people. These two parts interweave, criss-cross, and parallel, providing a play-within-play drama, contrasting then and now in 12 scenes, covering a span of 90 years: 1907 to 1997. The first part begins with the 1977 Hong Kong Repertory's rehearsal of Harriet Beecham's *Uncle Tom's Cabin*, traversing a setting between 1907's Tokyo and 1997's Hong Kong.

The second part opens with a 1997 interview of a local theatre director and crisscrosses the time and space between 1907 and the present. Both parts are thematically concerned with their respective quests, one purporting to perfect its artistry, the other to better the country and the people. The production relies technically more on devices of the screen and movies, with temporal and spatial contractions and ellipses, fragmentations, and episodes. The medium-istic mixture of this production reminds us of the experiment of the late British playwright, Joan Littlewood, in *Oh, What a Lovely War* in the 1960s. The use of slides and video interviews serves to narrate, explicate, and multiply the dramatic action of the play in such a way that the stage scene is not stuck with stasis and sluggish dialogue. These film clips' advantages preclude visual stasis and unnecessary dialogue. This production demonstrates the salient features of theatre imitating film, a sort of reversal of the trend at the start of the last century when cinema fed on theatre for form and content. However, this film-style feature can also be its weakness as the multimedia performance distracts and dilutes the content of the play at times. The audience would have liked to have the players stay with the scene longer, speak their minds and account for their movements more coherently, and develop their emotions more gradually. In other words, the theatre loses itself by not becoming more of itself. Most criticisms of this production attack its thinness of content as well as its loose form. All these may be attributed to the fact that the theatre is doing something else by playing its scenes in a more film-like fashion. This brings out an issue as to what constitutes the ontology and episteme of "liveness" in the theatre because of technological innovation or intrusion. This point is noticed by the famous German theatre scholar, Erika Fisher-Licht, in her article in *Modern Drama*, "*Quo Vadis?*: Theatre Studies at the Crossroads".[5] Live performance in the theatre with a growing mediatized format of representation defines the ontology of the here and now and makes audiences perceive "what boundaries there might be between the real and the imaginary, the exterior and interior of the performance".[6]

Another 2002 production by the same Drama Club is *Sunshine, the Station Master*, a play about a Chinese couple's enthusiasm for a life of sunshine and idealism in their search for China's ancient architecture and its preservation in the 1930s, a time of ideological collisions and the advent of the Japanese invasion. The production resorts to almost excessive use of archival film footage to tell the story and evoke the ambience of a bygone era, which are documentations of old architectural findings and life in wartime China. The many intermedial presences, lauded by the director, Ms Fu Yueh Mei, as facilitating fluid channels of narration and enriching "visual effects" on stage,[7] at times are distracting and often relegate the acting of the actor and the actress to a stop-and-go rhythm, leaving the dramatic fluidity and theatrical "liveness" fragmented and unstable, making the newly emerging aesthetics of the mediatization hard to sink into the audience.

These two productions display a very similar interest in Hong Kong's past, present, and future, together with a self-reflexive impulse towards the examination of the stage as a medium of exploration, representation, and even intervention, political as well as cultural. This fixation with the search for an ideal or identity has been a trademark of the Amity Club's productions since its inception.[8] It is only fitting for the issue of identity construction brought to public awareness both on the eve and in the aftermath of the 1997 handover.

The impact of media upon human behaviour is the main theme of another production, *Hamlet/Hamlet*, a Shakespearean adaptation by the joint venture of Amity Drama Club, Shatin Theatre Company, and Horizon Theatre Group, in 2001, featuring a provocative deconstruction of Shakespeare's *Hamlet*. The play is about a modern Hamlet contemplating his fate by watching 1948's Laurence Olivier playing Hamlet on a TV screen and plotting his revenge. The director, Xiong Yuan Wei (熊源偉), changed the original with a purpose, though retaining most of the original lines in Chinese. Television sets were installed on all sides of the rectangular stage, which was set in the centre of the theatre.

The incriminating evidence against Claudius for Hamlet's father's death, the revelation by the ghost, and the play-within-the-play episode took place on television screens. As if to compensate for the inevitable excision of the bard's language, the director used media technology as metaphors for illusion and reality, lies and truth, and determinacy and ambiguity and thematically raised the power of the media to the pedestal of an electronic God worship. When characters are faced with doubt and uncertainty, they look to the screen for guidance. In this play, stage reality intertextualizes with electronic reality on the TV screen, as modern Hamlet watches Laurence Oliver's Hamlet on the screen to find clues and justification for his revenge. Other "creative vandalisms" by the director include the ghost's appearance under the circumstances of a power failure, the electronic surveillance of Hamlet in place of Claudius and Polonius's eavesdropping, the substitution of the poetic death of Ophelia by her somewhat lurid death by electrocution, and the presence of a court photographer to record with a video cam all activities; all of these conspire to suggest the power of media and the intricate relationships of human behaviour.[9] Hence, mass media infiltrated the life of the characters and dictated their behaviours just as the TV screens mediatized the plot in this production. Man becomes enslaved to technology in a state of anxiety reminiscent of the British poet, W. H. Auden's prophetic announcement:

> ... this stupid world where
> Gadgets are gods and we go on talking
> Many about much, but remain alone,
> Alive but alone, belonging-where?[10]

The theme of media and communication branched out into the cyberspace of the Internet as January 2003 saw two Internet-inspired productions by the Hong Kong Repertory Company, *WWW.com* and *Rape Virus*. The former was written by a Shanghai playwright, Yu Jung-jun (于勇軍) and the latter by a local playwright, Chan Chi-wah (陳志華). *WWW.com* revolves around reality and illusion, communication and evasion of communication in a story in which an estranged couple plays out their desires and fantasies against the virtuality of online communication. The plot is concerned with how the estranged couple goes out to fulfil their desires and the happiness denied to each other in real life; the cyber-friendship and online confiding bestow on the couple new identities of intellectual and emotional fulfilment. The irony is that, to their shock and surprise, the new confidantes on the Internet turn out to be none other but themselves. The use of Internet technologies dramatizes the intriguing ways modern men communicate with one another or, rather, evade communicating with one another, a dilemma summed up by Harold Pinter in the 1960s:

> We have heard many times that tired grimy phrase,
> "failure of communication". I believe the
> contrary. I think that we communicate only too well
> in our silence; in what is unsaid, and that what takes
> place is a continual evasion, a desperate rearguard
> attempts to keep ourselves to ourselves. Communication
> is far too alarming. To enter into someone else's
> life is too frightening. To disclose to others the poverty
> within us is too fearsome a possibility.[11]

Rape Virus depicts the escapades of four newly acquainted line users' (Handicapped, Dick, Pussy, and SM) in the virtual world: plotting to rape a girl whom one of the four has secretly admired in a bizarrely imaginary and virtual manner through the Internet, bragging about sex and violence. With the help of the theatrical configurations of space, the play exploits the metaphoric rape virus and its epidemic irruption through the web. This production again highlights the theme of technologies and the human condition and the overdependence of humans on technologies. The computer virus affects the proper operation of not only the computer itself but the user as well. The characters in this play, so obsessed with ICQ communication or computer games and so consumed by their obsessions, are too incapacitated to distinguish between illusion and reality or between virtuality and real life; in a way, their freedom of action is predetermined by these technological gadgets, which take on a God-like potency to substitute anything that these characters may desire or fantasize and to affect their modes of sexual gratifications. The revealing name of one of the characters, Handicapped, symbolically epitomizes their existential conditions. Despite their fervent attempt at communication online, their existence remains one of incapacitation and alienation, a precise stage embodiment of Tennessee Williams's theme of interconnection and isolation:

> It is a lonely idea, a lonely condition, so terrifying
> To think of that we usually don't. And so we talk to
> Each other short and long-distance across land and
> Sea, clasp hands with each other at meeting and at
> Party, fight each other and even destroy each other
> Because of this somewhat always thwarted effort to
> Breakthrough the wall to each other.[12]

The technology-inspired productions share something in common, as they inevitably engender a new kind of aesthetics in the audience due to their form and content. This aesthetics is often computer-related with interactivity, immersive environments, and virtual reality, all of which can transform the audience's sense perceptions and create new forms of communication, fostering what Steve Johnson calls an "Interface culture."[13]

Innovation of the Theatre under Covid-19

Since the 2019s, havoc wreaked by Covid-19, creators of all art forms resorted more to the internet to let audiences enjoy online art and performance such as theatre, opera, concert, etc., just as churchgoers start to utilize its access to the AR/VR capabilities to transform the way faith and worship are conducted in the light of the Metaverse. These art forms, mixed with gamification devices such as avatars and the young generation's growing habituation to the virtuality of the Metaverse will strive to approximate or duplicate live experience with an added value, that is, inviting the audience's participation in instant exchange through a chatroom or in avatars, in the process of the performance, providing multiangle shots and surtitles, and by performing at venues of specific interest as well as shortening the playtime. All these experimental adaptations aim at arriving at a new way of theatre expression and interactive relationship with the audience under the circumstances.

On this note, I would cite two contemporary works, one in Hong Kong and another in London, as an endnote to the theme of this chapter: the integration of technology in art on the stage. Take, for example, the 2021 Production of *Du Fu 2.0* by the Prospect Theatre in Hong Kong. It is a stage production of one of the most famous Chinese poets of the T'ang Dynasty around 6th century. From beginning to end, the backdrop screen of the stage was utilized to project the Chinese character poetry of Du Fu to serve as a cinematic exposition and narration of his life and time. Poetry written in Chinese characters on the backdrop reinforces, with a montage-like juxtaposition, the dramatic action being enacted and visualizes the emotive content of the scene on stage, comparable to a technique of poetic fusion of scenery and emotion, *Ching Jing Jiao Yung* (情景交融) from classical Chinese poetry.

Another case in point is the Royal Shakespeare Company's 2021 online production of Shakespeare's *Midsummer Night's Dream*. A 50-minute drama inspired by the original, it blends live action with digital imagery and gaming technology. The play follows five of the sprites—Puck, Peaseblossom, Cobweb, Moth, and Mustardseed—through the woods, creating an online narrative. The theme of transformation in Shakespeare's play is embedded in the process. At every performance, the five actors in motion-capture suits shift around a physical space, sending their on-screen avatars moving around the virtual forest. During the performance, they react in real time with an online audience disguised as fireflies, inviting audiences worldwide to step into that mysterious forest and help shape the story. Audiences can choose to watch passively the story or interact with the actors/characters by becoming fireflies using their computer cursors to move around their screens at home. It is a piece of theatre plus a video game. The fusion of the physical and the virtual becomes a new way of storytelling.

These examples represent a new form of theatre to portray technology and human nature and offer an exciting theatrical synthesis of technology and art. But the mere presence of technology or mediatization cannot give birth to new viable and durable forms of dramatic art if the inclusion in the theatre of the technology of computers, the information highway, the World Wide Web, and even the three web toys is only a trendy veneer.

Live Performance and Mediatization

Luckily, most of the mediatized inclusion or incursion, as evidenced by the preceding productions, has done more than that. It is to be expected that this mediatized theatre will become more the bill of fare and integrated enough to transform the audience's consciousness or modes of perceptions, upon which the shaping of theatre will have to be based, so much so that the historical context of liveness in the expectations of the audience can accommodate and mitigate the usual bias that "the live event is 'real' and that mediatized events are secondary and somehow artificial reproductions of the real."[14]

There is no better time than here and now to raise a question concerning the issues of "liveness" and dramatic intensity in mediatized performance: Where is the line to be drawn so that this mediatization in the theatre will not overstep its bounds at the expense of the essence of the theatre as an art form? Perhaps the question could be asked from another angle: Is the line of difference between theatre performance and mediatization going to disappear somehow and somewhere in future? By asking these questions, I hope to acknowledge "the specificity to the experience of live performance",[15] as well as the ubiquitous presence of mediatization at the same time.

Perhaps we may even alternately accept the fact that the theatre has always been threatened by the "escalating dominance" of the media and that the theatre and other live forms of performance are competing dangerously with mediatized forms in terms of cultural formation and consumption. That said, I may sound as though I am begging the question of "a binary opposition" between the live and the mediatized, though this opposition has become more and more untenable, as the televisual and mediatized environment is fast becoming, in the opinion of Phillip Auslander, "an intrinsic and determining element of our cultural formation".[16] Theatre performance as live performance can lay claim to its liveness even in the ever-growing mediatized culture. Hence, in the light of the productions, theatre practitioners in Hong Kong, like their counterparts in the rest of the world, should recognise the relationship between the theatre and mediatization, diminish their unnecessary anxiety about the erosion of live performance, and instead realize the growing blur between the live and the mediatized at the aesthetic and epistemological levels in the age of the Metaverse.

Notes

1 This chapter is produced mainly from the *International Journal of Techno-Humanities*, Volume 1, No.1, pp. 35–43, with the kind permission of the journal editor.
2 Arnold Aronson (2009). "Technology and Dramaturgical Development: Five Observations", *Theatre Research International* 24(2): 188–97.
3 A brainchild of Joshua Putnam Peskay, the writer; Jessica Putnam Peskay, an actress in the play who is married to Mr. Peskay; and Mathew A. Peskay, the director and Mr. Peskay's brother. The play is about unraveling some of the complexities of the Internet and technology.
4 Wave 2003, an experimental group, staged two performances this year, *2056-Pilot Version 1.0* (2056–測試版 1.0), *A Simple Day in the Life of Paradox* (柏羅托斯簡單的一天); other examples include Zuni Icosahedron's *Lover's Discourse* and Hong Kong Repertory Company's *Formula of the Thunderstorm* (雷雨謊情), David Lam Theatre's *Happy Prince* (非常林奕華) (快樂王子). All of these works incorporated technology as form or content in their stage productions.
5 Erika Fisher-Licht "*Quo Vadis?*: Theatre Studies at the Crossroads," *Modern Drama*, 44:1 (Spring 2001): 60–63.

6 Alice Rayner, "Everywhere and Nowhere: Theatre in Cyberspace," in *Of Borders and Thresholds: Theatre History, Practice, and Theory*, edited by Michael Kobialka. Minnesota: University of Minnesota Press, 1999, p. 278.
7 See notes by the director in the "Sunshine, Station Master" Programme, 2002.
8 Previous works of the Amity Drama Club, *The Clan of Generals* (1981), *Man of La Mancha* (1982), *Man! Oh, Man* (1985), *The Search for the Spring Willow* (1997), and *the Boat Sail on* (1997) have all borne this trademark.
9 Xiong Yuan Wei (2000). "From Classic to Modern: A Director's Note to Hamlet/Hamlet", *Hong Kong Drama Review* 2: 93–94.
10 W. H. Auden (1947) *The Age of Anxiety*, New York: Random House, p. 44.
11 Harold Pinter (1961). "Writing for Myself", *Twentieth Century*, (February): 172–75.
12 Tennessee Williams (1951). "Preface to Cat on a Hot Tin Roof", New York: New Directions, p. vii.
13 Andrew Murphie and John Potts (2003). *Culture and Technology*, London: Palgrave-McMillan, p. 86.
14 Philip Auslander (1999). *Liveness: Performance in a Mediatized Culture*, London and New York: Routledge, 3.
15 Ibid.
16 Ibid., 2.

References

Aronson, Arnold (2009) "Technology and Dramaturgical Development: Five Observations", *Theatre Research International* 24(2): 188–97.
Auden, W. H. (1947) *The Age of Anxiety*, New York: Random House.
Auslander, Philip (1999) *Liveness: Performance in a Mediatized Culture*, London and New York, Routledge.
Fisher-Licht, Erika (2001) "*Quo Vadis?*:Theatre Studies at the Crossroads", *Modern Drama* 44(1): 60–63.
Murphie, Andrew and John Potts (2003) *Culture & Technology*, London: Palgrave McMillan.
Pinter, Harold (1961) "Writing for Myself", *Twentieth Century*, February, 172–75.
Rayner, Alice (1999) "Everywhere and Nowhere: Theatre in Cyberspace," in Michael Kobialka (ed.) *Of Borders and Thresholds: Theatre History, Practice, and Theory*, Minnesota: University of Minnesota Press, 278–302.
Xiong, Yuan Wei 2000) "From Classic to Modern: A Director's Note to *Hamlet/Hamlet*", *Hong Kong Drama Review* 2: 93–94.

3 Densities and Fugitive Vectors

Grant Hamilton

The nascent condition of what one might call "techno-humanities" perhaps explains the fact that it remains undertheorized. However, in an early position piece, Mikhail Epstein outlines what I think is a productive way to begin thinking about the contours of the term (2006). In brief, and in distinction to the concerns of the traditional humanities,

> techno-humanities explore that which does not yet exist. They project and produce possible cultural objects and forms of activity including new artistic and intellectual movements, new disciplines, research methodologies and philosophical systems, new styles of behavior, social rituals, semiotic codes, and intellectual trends.
>
> (Epstein 2006)

At the core of this pronouncement is the assertion that there is a need for a practical branch of humanities that is conscious of the necessity to build "new intellectual communities, new paradigms of thinking and modes of communication rather than simply studying or criticizing the products of culture" (2006). The watchword here is "transformation"—that is, techno-humanities is that which is explicitly concerned with producing work that affects "new cultural positions, movements, perspectives, and modes of reflexivity" (2006). In other words, techno-humanities is a field of study that is interested in reclaiming and reasserting the *techne* of cultural enquiry and, in this way, bringing art, skill, and creativity closer to the humanities. As such, while one might immediately associate the term with the integration of technology into the practice of the humanities, it is not necessary to do so. But neither is it sensible to summarily dismiss the employ of technology from such enquiry, since advances in technology are often a boon to the creativity that the techno-humanities seeks to foster.

Indeed, this is the proposition that underscores this essay—that a computational approach to literary studies has the potential to both expand and enrich our shared appreciation of literature. Unfortunately, the relationship between computational literary studies and more traditional critical approaches to

DOI: 10.4324/9781003376491-4

literature has been somewhat combative (see Da 2019). Yet it is clear that work in this field has produced results of great value to the literary studies community as a whole—and continues to do so. By asking very different questions of texts to those typically posed by the traditionally schooled literary scholar, critics such as Franco Moretti (2005, 2013), Matthew Jockers (2013), and Ted Underwood (2019) have certainly extended the current understanding of Western literary history. However, my own interest in the field emerges not from the analysis of the grand movements of literature across the centuries but from the way in which such computational approaches to literary studies allow the critic to better understand the singular book. More specifically, I am interested in how computational approaches allow the critic for the first time to truly understand and explore the image of the book as discussed by the French philosophers Gilles Deleuze and Félix Guattari some 40 years ago. At the beginning of *A Thousand Plateaus* (French 1980; English 1987), Deleuze and Guattari write:

> A book has neither object nor subject; it is made of variously formed matters, and very different dates and speeds. To attribute the book to a subject is to overlook this working of matters, and the exteriority of their relations. It is to fabricate a beneficent God to explain geological movements. In a book, as in all things, there are lines of articulation or segmentarity, strata and territories; but also lines of flight, movements of deterritorialization and destratification. Comparative rates of flow on these lines produce phenomena of relative slowness and viscosity, or, on the contrary, of acceleration and rupture. All this, lines and measurable speeds, constitutes an assemblage. A book is an assemblage of this kind, and as such is unattributable.
> (Deleuze and Guattari 1987, 3–4)

To one degree or another, my career as a literary critic has been spent trying to gain a full understanding of the dynamics that Deleuze and Guattari outline in this passage. Only now, though, with the advent of a technology that allows one to accurately parse a text into vector space do I feel as though this Deleuzo–Guattarian image of the book can be fully appreciated and finally employed as a means to critically interrogate a text. Upon penetrating the characteristically dense language of Deleuzo–Guattarian thought, the critic learns here that no book is of a singular form or a singular content. That is to say, every book is dynamic in its constitution and in terms of the way that the "meaning" of a book is dependent on the ever changing encounter of the reader and text. But having said that, it is also clear that Deleuze and Guattari postulate a structure to a book—a "geology", as they would have it. The difficulty has always been how best to reconcile the image of the book as that which boasts an ever changing structure.

My conjecture here is that a productive way to begin to understand the geology of the text as Deleuze and Guattari write it is to think of words as vectors—vectors that at times cluster together to form densities that gesture towards meaning, while at other times escape from one another to such an extent that one is forced to consider the significance of the outlier. Indeed, this is what it means to think of books in terms of "dates and speeds". It is an invitation to consider texts in terms of the movement of convergence and divergence, similarity and dissimilarity, or, put otherwise, clusters and outliers. For the first time, the literary critic has the ability to articulate what one might think of as the densities and eccentricities of a text with an accuracy that comes from statistical analysis. And this is what this chapter seeks to articulate: the densities and fugitive vectors of a text. To do so is not only to begin to elucidate the fundamental structure of a text but also to step towards what Franco Moretti has called "a falsifiable criticism" (1983, 23)—a literary criticism that rests on statistical models and results that can be objectively tested and objectively repeated (therefore rendering the criticism, as a whole, falsifiable).[1] This is not to be mistaken for a move to "scientize" literary criticism or, more broadly, the humanities[2] but rather to embrace true experimentation and, for that, creativity.[3] The text upon which I experiment here is J. M. Coetzee's well-read novel, *Waiting for the Barbarians* (1980).

Perhaps the most obvious way to begin a computational analysis of Coetzee's novel is to employ a simple count instruction. Once common "function" words have been removed from the count (that is, "stopwords" such as *the*, *and*, *of*, and so on),[4] one can begin to get a clearer sense of the most commonly used words in *Waiting for the Barbarians*.

Although this produces an interesting list of words, such a raw word-frequency list is at best suggestive of the concerns of the text under analysis.

Table 3.1 Frequency of the 15 Most Common Words in *Waiting for the Barbarians*

Feature	Frequency
say	314
one	253
man	244
can	223
back	219
see	210
come	190
like	177
day	172
go	168
know	154
hand	149
barbarian	141
eye	125
time	124

Note: This raw word-frequency list has been produced after stopwords have been removed and after the text has been lemmatized (i.e. *men* and *man* counted under *man*). On the lemmatization of texts, see Benoit and Matsuo (2020).

One might speculate from browsing this list of terms that Coetzee's novel is in some way engaged with the observation (*see/eye*) and reporting (*say*) of personal experiences and that barbarians (*barbarian*) are important conceptual figures of the text, but beyond this, little more can be inferred. The problem in generating such lists is twofold. First, the most significant words of a literary text are not always the most frequently written (otherwise the stopwords that are typically removed from such lists would always prove to be the most significant words of a text);[5] second, words taken in isolation can only reveal so much information about the text they populate. As the noted linguist J. R. Firth memorably quipped: "You shall know a word by the company it keeps!" (1962, 11). Thus understanding how a particular word is used in association with other words in a text may reveal something of wider significance to the text at large.

With this in mind, one can begin to investigate the relationship between words in a text as a means of identifying clusters or densities of significance. One way to begin such an investigation is to interrogate the co-occurrence of words. If one maps the co-occurrence of words in *Waiting for Barbarians*, then the image shown in Figure 3.1 emerges.

This image certainly gives one an impression of how the co-occurrence of words in Coetzee's novel begins to form a density of sorts, but this density lacks the definition necessary to extend the critic's appreciation of *Waiting for the Barbarians*. A rather more nuanced metric to investigate the relationship of words to a text, and one that also takes into account the relative triviality of ultra-high-frequency words, is called a "term frequency–inverse document frequency" (TF-IDF) weighting.[6] While originally conceptualized as a means of evaluating the statistical significance of a word across a corpus of texts, TF-IDF calculations can also be made to show the importance of a particular word to a specific text. By artificially creating a "corpus" through the partitioning of a single text into multiple sections, one can productively employ the TF-IDF algorithm. Table 3.2 gives those terms which emerge as (most to least) statistically significant to the novel as a whole as one changes the size by which the text is divided.

There are a number of points to note about these results. First is the appearance of a number of new words to those found in the raw word-frequency list (see Table 3.1). In addition to words such as "foot", "room", and "boy", the TF-IDF weighting of the text reveals the persistent significance of "horse" and "prisoner" to the text regardless of the size of the sections into which the text has been partitioned. To those who are familiar with the novel, it is perhaps unsurprising that the term "prisoner" assumes such a position of statistical significance. After all, the conceptual and substantive figure of the prisoner dominates Coetzee's novel. From the unnamed prisoner who dies at the hands of his interrogators at the beginning of the novel to the Magistrate who becomes a prisoner after submitting to a fit of conscience, *Waiting for the Barbarians* can profitably be considered an examination of the ethics, affects, and experiences of the prisoner. Even so, it is interesting

32 *Grant Hamilton*

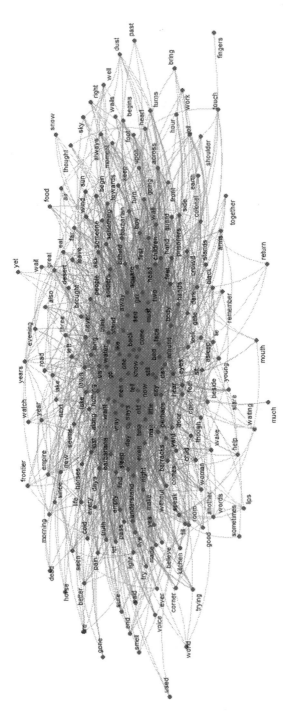

Figure 3.1 Feature co-occurrence network of *Waiting for the Barbarians*.

Note: This feature co-occurrence network was calculated using the Quanteda package in R. See Benoit et al (2018). It shows the relationship between the 200 most frequent words in Coetzee's *Waiting for the Barbarians*. The frequency count threshold for the calculation was 0.85, and those features which have not met this threshold have been omitted from the network.

Table 3.2 Top TF-IDF Terms for *Waiting for the Barbarians* by (Changing) Partition Size

Definition (words)	Top-weighted features of novel (10)
200	barbarian, horse, prisoner, tell, we, foot, body, sleep, say, want
300	horse, prisoner, barbarian, boy, foot, we, sleep, room, tell, water
400	horse, prisoner, boy, room, wind, guard, yard, tent, sleep, water
500	horse, prisoner, boy, guard, room, door, tent, crowd, colonel, truth
1000	horse, tent, prisoner, mandel, room, crowd, sand, rope, girl, guard

Note: In this example, the text has been lemmatized before being partitioned into 200-word sections (producing 153 sections in total), 300-word sections (producing 102 sections in total), and so on. While the results from the novel partitioned into 200-word sections can be said to be of a higher definition than the novel partitioned into 1,000-word sections, one should not infer that results from a 200-word partition are in some way "better" than results from a 1,000-word definition. What this table shows is that different words come to the fore as the "resolution" of examination changes.

that the TF-IDF weighting algorithm raises this term from 66th on the raw word-frequency list to such a position of prominence.[7] The prominence of the word "horse" is certainly more unanticipated than "prisoner", and in lifting the term from 52nd on the raw word-frequency list to the very top of the weighted features list, the TF-IDF algorithm forces the critic to think again about the significance of the image of the horse in Coetzee's novel. In total, the term appears 80 times in the text. Upon closer inspection, the figure of the horse variously and simultaneously symbolizes, represents, or simply is property to be protected, stolen, or traded, a form of transport (willing or not), a beast of burden or labour, inextricable from the figure of the nomad/barbarian who, aside from the crippled barbarian girl, is never seen dismounted from the animal[8] and, perhaps most interestingly of all, is seen as evidence of human non-exceptionalism.[9] Clearly, there is much to be said about this figure in the novel—not least that a careful reading of the figure of the horse unveils the complex way in which Coetzee configures the moral and ethical aspects of human interaction with the non-human world (an abiding interest that will constantly emerge in Coetzee's writing) in the text—yet it seems as though the horse has been roundly ignored in the scholarship on the novel.

One important way in which the critic can begin to understand the idea or image of the horse and how it functions in this particular novel is to inspect a word embedding algorithm. Typically, a word embedding algorithm is an unsupervised mechanism by which one can obtain vector representations for the words of a given text. In Table 3.3, the embedding algorithm GloVe (Pennington et al. 2014) has been used to encode the ratios of word–word co-occurrence probabilities as a means to identify a group of words that play a significant role in building the specific meaning of a word as it takes shape in the text.[10]

Table 3.3 Nearest Neighbour for Top-rated TF-IDF Terms in *Waiting for the Barbarians*

Feature	Nearest Neighbour Terms (10) in Order of Proximity (Nearest to Farthest)
horse	take, sky, stopped, papers, skull, ways, talk, front, apartment, sits
prisoner	became, instant, easy, plant, remember, harm, surface, shouts, anger, hear
barbarian	among, know, prisoners, friends, raid, remains, girl, strange, pick, covered
boy	little, barbarian, ask, pain, just, man, lies, sick, stand, breast, difficult
foot	show, plan, history, another, year, pleasure, streets, call, instant, asks

Such densities that form around terms will either be in sympathy with the language that literary critics have employed to discuss such features of a novel, or they will (preferably) provoke consideration of new relationships. For example, in addition to the common acts of taking (*take*) and sitting (*sits*) on horses, the cluster that forms around "horse" here also compels one to reflect on the relationship between "horse" and "sky" as it evolves in Coetzee's novel. Indeed, once alerted to this association, it becomes clear that this is a relationship of some degree of significance.

For the unnamed Magistrate, the sky is a substance of sorts—the kin of aether (luminiferous aether) but also extended so that it functions as the threshold by which things come in and out of being. While the sky holds the natural world in all its vast and terrifying glory (stars, trees, birds, deserts, dust storms, and so on), it is also that which births the presence of the unknown. The Magistrate's first encounter with the so-called barbarian people who live beyond the gates of his frontier town evidences the point. They appear as "twelve mounted men on the skyline" (94). A little later, the Magistrate repeats the association: "the barbarians stand outlined against the sky above us" (96). Given the way that the Magistrate previously describes the sky as "dissolving" and "blending" the edges of material reality,[11] it is as if the reader is invited to regard this encounter as almost spectral in nature—the barbarian figures emerging into a material world from an unknown elsewhere. If understood like this, then the typically unremarked horses upon which the barbarians *always* sit assume something like a mythical bearing. Here, horses are creatures that have the capacity to transition between worlds (between the worlds of the known and unknown) and, for that, are creatures worthy of profound veneration and awe. Perhaps it is in recognition of the treatment of horses in the world of Empire—as objects to be traded, as beasts of burden, as creatures forced into unsuitable terrain to the detriment of their health—that the barbarians are forced to demand a horse when negotiating the safe passage of the barbarian girl with the Magistrate. Curious for the fact that the horse demanded was clearly in what the Magistrate calls "a bad way" (98), it is almost as if the barbarians see an opportunity to rescue an empyrean being from the tortures of the human world.

In this way, exploring the densities which emerge around terms following the implementation of a word embedding algorithm like GloVe has the potential to highlight previously unexplored relationships within a novel. Importantly, such unmasking of previously unregarded moments in a text persists if one moves critical attention away from word co-occurrences and focuses instead on what one might call the sectionality of a novel. So, if one continues with the partitioned novel required for the TF-IDF weighting of terms, then it is possible to note the emergence of another kind of density—a density premised on the semantic relationship between sections of the novel, which can be visualized as in Figure 3.2.

An interrogation of the data that produces this network shows that the most "connected" sections in Coetzee's *Waiting for the Barbarians* (which is to say, those 300-word sections that are related semantically to the most other sections of the novel) are sections 99 and 100.[12] Similarly, the least connected section (or the outlier section of the text) is section 53.[13] Clearly, these sections are important to the novel in one way or another, but it is perhaps where these sections sit within the text at large that discloses their significance. Indeed, such an observation might suggest that Coetzee's novel is structured along the lines of a classical (Aristotelian) tragedy. It seems reasonable to suggest that sections 99 and 100 are so semantically integral to the novel because they serve as the denouement or the moment in which the various lines of interest opened up by the plot come to some sort of resolution in the text—none more apposite than Mai's blunt declaration to the Magistrate that, regardless of the world that he thought he had engineered around the barbarian girl, he had in fact made her "very unhappy" (203).

But understanding the significance of the outlier section 53 to the novel requires a little more elaboration. Mapping the emotional valence of Coetzee's novel shows that *Waiting for the Barbarians* is an emotionally challenging text. Beginning from a neutral position, a cumulative sum of the emotional valence of the novel shows an irremediable march towards the negative. See Figure 3.3.

Of course, this tallies well with a text that is explicitly concerned with State-sanctioned torture and the anxiety of being on the frontier of a civilization that is seemingly imperilled by an imminent and violent threat. However, a closer inspection of the data reveals a degree of nuance in this trajectory towards desolation. Directly following section 53, the descent into despair accelerates, and so much so that this section can be regarded as a key transitionary moment in the novel. Concerned with the figure of a father who realises that he cannot stop the beating of his own child at the hands of the State, the death of affect suffered by the child who recognizes the impotence of such a father, the Magistrate who continues to prey on a young girl in full knowledge of such events and his own behaviour, as well as the realization

36 *Grant Hamilton*

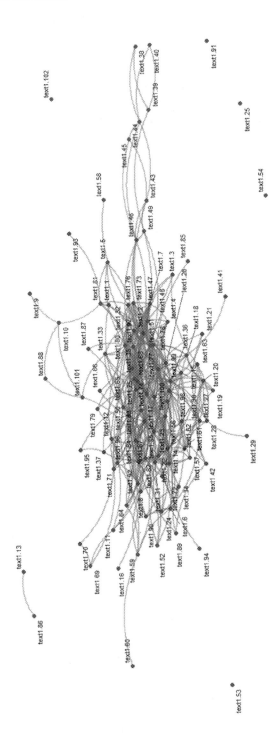

Figure 3.2 Section network of Coetzee's *Waiting for the Barbarians*.

Note: This network visualizes a distance matrix that is generated from partitioning Coetzee's novel into 300-word sections and then coercing the result into a document feature matrix. See Benoit et al (2018).

Densities and Fugitive Vectors 37

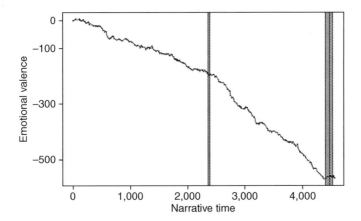

Figure 3.3 Graph showing the cumulative sum of the emotional valence of *Waiting for the Barbarians* and the position of sections 53, 99, and 100.

Note: The emotional valence of the text was calculated using the Syuzhet package in R. See Jokers (2015).

that the Empire is working to destroy itself as it searches for ways to nullify the threat of the supposed barbarian forces that lie just beyond the horizon, section 53 can be read as the moment of "recognition" (or *anagnorisis*) that inaugurates the climax and ultimate reversal of the tragic plot. See Figure 3.4. Given this, it is perhaps unsurprising to learn that Ato Quayson chooses to discuss precisely this novel in his book on *Tragedy and Postcolonial Literature* (2021).

Even though an analysis of the section network of *Waiting for Barbarians* begins to render visible the plot structure of the novel, it is important to note that such an analysis does not describe the full and final architecture of the novel. It is true that an almost untold number of word clusters transit the text—word clusters that disclose yet more densities within the novel. To get a sense of the nature of such densities, one need only submit the text to a topic modelling algorithm. Such an algorithm is capable of highlighting latent semantic relationships between words and, in this way, identify the word clusters, or "topics", that traverse a text. Although rarely coherent in the way that most classically trained literary critics talk of topics in literature, such algorithmic topics nonetheless have the potential to reveal previously unacknowledged textual relationships. Table 3.4 shows the ten most coherent topics—that is, word clusters with the most stable set of features—that appear in *Waiting for the Barbarians*. Interestingly, the labels given to each topic here are extrapolated and assigned by the topic modelling algorithm itself.[14]

Perhaps best thought of as "tight densities", highly coherent topics are clusters of words that are linguistically bound together in the text but do

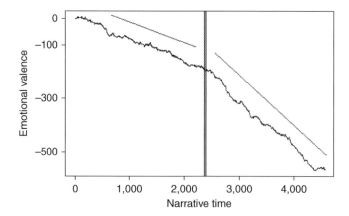

Figure 3.4 Graph showing the cumulative sum of the emotional valence of *Waiting for the Barbarians* and the change in frequency of negative emotional language that occurs at section 53.

Table 3.4 Top Topics by Coherence in *Waiting for the Barbarians*

Topic	Native Topic Label	Coherence	Top Terms in Topic
119	warrant_officer	0.413	officer, warrant_officer, warrant, body, joll, brother, sheet, barbarian, character, yesterday
35	aroused_daydreaming	0.385	marrows, ugly, table, heavy, eyes, abandon_locutions, abandon_regretfully, abandoned_afford, abandoned_gatepost, abandoned_suppose
89	shit_shit	0.385	shit, hear, shit_shit, roof, dont, fucking, mat, hut, friends, dust
95	amid_flying	0.324	square, roar, volleys, stands, fusillade, dismount, shots, flowers, steady, nearer
68	live_live	0.290	rope, live, ladder, live_live, playing, climb, rung, bag, holds, arm

Note: This table shows the nearest neighbours for the top-rated TF-IDF terms in *Waiting for the Barbarians* after the text has been partitioned into 300-word sections.

not necessarily repeat through the text at large. Typically, these are groups of words that belong to a specific passage (or two) in the text, as can be seen from the graph in Figure 3.5, which shows when and to what extent (known as the theta score) topic 119 ("warrant_officer") occurs throughout Coetzee's novel.

Although the critic can detect a weak signal of this topic towards the middle of the novel, it is clear from the graph that the cluster of words that describe

Densities and Fugitive Vectors 39

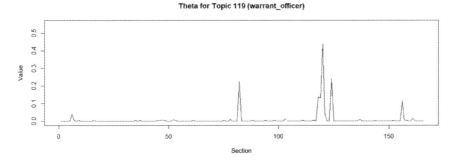

Figure 3.5 Graph showing the distribution of topic 119 (warrant officer) in *Waiting for the Barbarians*.

Table 3.5 Top Topics by Prevalence in *Waiting for the Barbarians*

Topic	Native Topic Label	Prevalence	Top Terms in Topic
10	day_day	1.039	dont, arm, understand, told, day, barbarians, cry, talk, wait, day_day
37	begin_wash	1.003	feet, foot, wash, basin, legs, begin, warm, begin_wash, head, bandages
124	telling_truth	1.001	oy, guard, sore, listen, lantern, blood, speak, excellency, tied, shut
119	warrant_officer	0.970	officer, warrant_officer, warrant, body, joll, brother, sheet, barbarian, character, yesterday
40	prisoners_questioned	0.961	prisoners, time, questioned, hall, happened, corner, daughter, duty, hurry, sick

topic 119 predominantly belong to a scene in which the Magistrate imaginatively translates for the paranoid and mistrustful Colonel Joll the enigmatic poplar slips that he has retrieved from the desert (149–151). The story he invents for Joll is one that echoes the opening scenes of the novel—a barbarian father travels to the frontier town in order to reclaim the body of his son who, the Magistrate implies, died at the hands of his imperial interrogators. The signal of this topic is of such strength that it marks a density in the text that the attentive reader would perhaps be unwise to dismiss too quickly.

In distinction to this, there are topics that are significant for the fact that they are more prevalent in a text than those which form "tight densities". Indeed, it is perhaps sensible for one to think of such prevalent topics as reflecting the abiding interests of a text—that is to say, those aspects of a text that are constantly put in front of the reader. In *Waiting for the Barbarians*, the most prevalent topics are shown in Table 3.5.

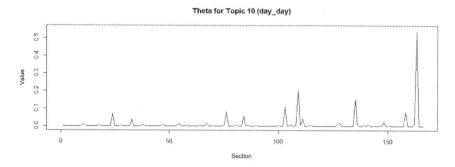

Figure 3.6 Graph showing the distribution of topic 10 (day_day) in *Waiting for the Barbarians*.

The cluster of words that constitute topic 10 assumes greatest coherence in a passage towards the end of the novel where the Magistrate and Mai share their apprehensions about the imminent arrival of the barbarian hordes at the gates of their undefended frontier town (202–203). Here, they are literally waiting for the barbarians. If there is a prevalent and dominant image of the novel, then statistically this is it. Significantly, such "recursive densities" as this might lack the coherence of other topics (although not always—topic 119 appears on both lists), but they certainly represent a persistent aspect of the novel. Indeed, mapping the theta of the most prevalent topic in Coetzee's novel makes clear just how persistent the topic of waiting for the barbarians is in the novel at large. See Figure 3.6.

Read together with the other prevalent topics selected by the topic modelling algorithm, one might reasonably conclude that the most insistent images of Coetzee's novel are those of people who are in some fashion anticipating the arrival of a devastating unknown force ("day_day"), the Magistrate's ceremonial washing of the barbarian girl's feet ("begin_wash"), the nature of truth telling ("telling_truth"), the uncomfortable nature of truth told to power ("warrant_officer"), and the poorly constructed stories told by those in authority to cover over moments of "indiscretion" ("prisoners_questioned"). Few critics, I think, would disagree with the significance of such a catalogue to understanding the concerns of the novel.

That computational literary analysis has the capacity to both affirm and expose avenues of engagement with literary texts in this way is its great strength. As has been demonstrated in this chapter, as a methodology, computational literary studies is able to highlight those moments of a text in which vectors either flee from one another or otherwise form various kinds of densities—densities that have the capacity to (re)direct a reader's attention to a particular image, a particular cluster of words, or a particular passage of a text, and in this way potentially mark out previously unremarked areas for further critical interest. Given this, what some may view as a threat to the discipline of literary criticism is for others the chance to find novel ways of thinking about and discussing literary texts (whether well-read or not). In this light, the profit of computational literary analysis is not to be found in the way that it rivals the work of the traditional literary critic but rather in the way that it aids the act of criticism.

The algorithms that seem to "discover" the aspects of the text that one can readily find in the critical literature are the same algorithms that produce curious, enigmatic, or sometimes just plain strange results. Perhaps there are fundamental yet unexplored connections to be found amongst such results; perhaps there are not. There is little profit to be made from the suggestion that the algorithms used in this chapter are flawless—statistically they are sound, of course, but the application of them may be found by some to be at times a little wanting. It is hard to defend against the accusation that such analysis is still raw and, for that, tentative. But equally, such analysis is at the same time indicative of a new and exciting way to interrogate literary texts. To those who adopt such a hybrid mode of criticism, where calculations highlight statistically important words, associations, and passages for the literary critic to pursue, the reward is a re-enlivened world of literature. Everything must be read again. And in this grand rereading of our shared literary heritage, new intellectual communities and new paradigms of thinking will surely emerge. Such is the vision of literary studies within the techno-humanities.

Notes

1 In the spirit of such falsifiable criticism, it should be noted that the code used to generate the findings of this chapter is available for public scrutiny and use at www.onworldliterature.wordpress.com. All calculations were processed using the R language and environment for statistical computing. See R Core Team (2020). All packages used in the calculation of the data for this chapter are referenced in the works cited of this chapter. Finally, it should be noted that the graphic visualizations of data given here were constructed either through base R functions or the ggplot package in R. See Wickham (2016).
2 Mikhail Epstein has also cautioned against this move (2006).
3 One of the major refrains of Deleuze and Guattari's *A Thousand Plateaus* is "experiment, don't interpret!" See, for example, Deleuze and Guattari (1987, 139).
4 On the significance of removing stopwords, see Gerlach et al. (2019). In short, stopwords can be thought of as those high-frequency "function" words—such as articles, common adverbs, conjunctions, pronouns, and prepositions—that lend little thematic weight to a text. See Jockers (2013, 131).
5 It is perhaps worth stressing the point. While the top feature reported in Figure 3.1 is "say" (which occurs 314 times in Coetzee's novel), the most common word used in the text is "the" (which occurs 4,743 times). Indeed, the frequency of stopwords accounts for just over 59% of the total frequency of all words in the novel.
6 On TF-IDF, see Manning et al. (2008).

7 The rank given here is after the removal of stopwords and the lemmatization of the text. In terms of raw count, the word "prisoner" is used 66 times in Coetzee's novel.
8 This observation alone will be enough to provoke some to examine the affect(s) of a Deleuzian nomad/barbarian becoming-horse. See Deleuze and Guattari (1987, 232–309).
9 For example, this is precisely the way in which the Magistrate uses the figure of the horse when imagining the competing epistemology of the barbarian girl, whom he has nefariously brought into his chamber: "I prefer not to dwell on the possibility that what a barbarian upbringing teaches a girl may be not to accommodate a man's every whim, including the whim of neglect, but to see sexual passion, whether in horse or goat or man or woman, as a simple fact of life with the clearest of means and the clearest of ends" (76).
10 This calculation was run on the Text2Vec package in R. See Selivanov et al. (2020).
11 Early in the novel, the Magistrate relates a dreamscape: "From horizon to horizon the earth is white with snow. It falls from a sky in which the source of light is diffuse and everywhere present, as though the sun has dissolved into mist, become an aura. In the dream I pass through the barracks gate, pass the bare flagpole. The square extends before me, blending at its edges into the luminous sky" (15).
12 It is perhaps worth restating here that, if partitioned into 300-word sections after the removal of stopwords and lemmatization, *Waiting for the Barbarians* boasts a total of 102 sections. Those interested in reading sections 99–100 should refer to pages 199–204 of the edition of the novel cited in the References.
13 Similarly, section 53 corresponds approximately to pages 109–111.
14 Within topic modelling algorithms, the *k* value determines the number of topics that the modelling algorithm attempts to populate. While the *k* value is usually a supervised part of this algorithm (that is, set by the analyst), the *k* value used in this example is also selected by algorithm. For an unsupervised way of determining *k*, see Jones (2019b). The topic modelling package used here was TextmineR. See Jones (2019a).

References

Benoit, Kenneth and Akitaka Matsuo (2020) *spacyr: Wrapper to the 'spaCy' 'NLP' Library*. R package version 1.2.1, available from https://CRAN.R-project.org/package=spacyr

Benoit, Kenneth, Kohei Watanabe, Haiyan Wang, Paul Nulty, Adam Obeng, Stefan Müller, and Akitaka Matsuo (2018) "Quanteda: An R package for the quantitative analysis of textual data", *Journal of Open Source Software* 3(30): 774, available from https://quanteda.io

Coetzee, J.M. (2000) *Waiting for the Barbarians*, London: Vintage.

Da, Nan Z. (2019) "The Computational Case against Computational Literary Studies", *Critical Inquiry* 45(3) (Spring): 601–39.

Deleuze, Gilles and Félix Guattari (1987) *A Thousand Plateaus: Capitalism and Schizophrenia II*, translated by Brian Massumi, Minneapolis: University of Minnesota Press.

Epstein, Mikhail (2006) "Towards the Techno-Humanities: A Manifesto", *Art Margins* (13 Mar), available from https://artmargins.com/towards-the-techno-humanities-a-manifesto/

Firth, J. R. (1962) "A Synopsis of Linguistic Theory, 1930–1955", in J.R. Firth (ed.) *Studies in Linguistic Analysis*, Oxford: Blackwell, 1–32.

Gerlach, Martin, Hanyu Shi, and Luís A. Nunes Amaral (2019) "A universal information theoretic approach to the identification of stopwords", *Nature Machine Intelligence* 1: 606–12.

Jockers, Matthew (2013) *Macroanalysis: Digital Methods and Literary History*, Chicago: University of Illinois Press.

Jockers, Matthew (2015) *Syuzhet: Extract Sentiment and Plot Arcs from Text*, available from https://github.com/mjockers/syuzhet

Jones, Thomas (2019a) *textmineR: Functions for Text Mining and Topic Modeling*, R package version 3.0.4, available from https://CRAN.R-project.org/package=textmineR

Jones, Thomas (2019b) "3. Topic Modeling", available from https://cran.r-project.org/web/packages/textmineR/vignettes/c_topic_modeling.html

Manning, Christopher D., Prabhakar Raghavan, and Hinrich Schütze (2008) *Introduction to Information Retrieval*, Cambridge: Cambridge University Press.

Moretti, Franco (2013) *Distant Reading*, London: Verso.

Moretti, Franco (2005) *Graphs, Maps, Trees: Abstract Models for a Literary Theory*, London: Verso.

Moretti, Franco (1983) *Signs Taken for Wonders: On the Sociology of Literary Forms*, London: Verso.

Pennington, Jeffrey, Richard Socher, and Christopher D. Manning (2014) "GloVe: Global Vectors for Word Representation", in *Proceedings of the 2014 Conference on Empirical Methods in Natural Language Processing*, 1532–1543, Doha: Association for Computational Linguistics.

Quayson, Ato (2021) *Tragedy and Postcolonial Literature*, Cambridge: Cambridge University Press.

R Core Team (2020) *R: A language and environment for statistical computing. R Foundation for Statistical Computing*, available from www.R-project.org/

Selivanov, Dmitriy, Manuel Bickel, and Qing Wang (2020) *text2vec: Modern Text Mining Framework for R*. R package version 0.6, available from https://CRAN.R-project.org/package=text2vec

Underwood, Ted (2019) *Distant Horizons: Digital Evidence and Literary Change*, Chicago: University of Chicago Press.

Wickham, Hadley (2016) *ggplot2: Elegant Graphics for Data Analysis*, New York: Springer-Verlag.

Wickham, Hadley (2021) *tidyr: Tidy Messy Data*, R package version 1.1.3., available from http://CRAN.R-project.org/package=tidyr

4 Revisiting the Future of Translation Technology

Chan Sin-wai

Introduction

Translation technology has been fast-changing, so fast that a few years after the publication of my book *The Future of Translation Technology: Towards a World without Babel* (Routledge 2017), I find it necessary to revisit my pet theme by re-examining some of the major issues in translation technology.

In the following, we will discuss the name and nature of translation technology; study the developments of different forms of translation; analyse the advantages and disadvantages of using machine translation, computer-aided translation, localization, and speech translation; and look at the future of these technologies. In the Conclusion, we will look into the future of translation technology in a holistic manner.

Name and Nature of Translation Technology

What is translation technology? In recent years, we talk a great deal about translation technology. TranTech is on the lips of all translators and interpreters. But most people do not seem to have a clear idea of the name and nature of translation technology and what it is about.

The name of translation technology has to do with the definitions of TranTech. To date, there are two major definitions of translation technology: One is given by Lynne Bowker of the University of Ottawa, and the other by me.

According to Bowker, translation technology is about the use of tools in translation practice (Bowker 2002: 5–9). Tools can be general or specific. In translation practice, both are needed. General tools refer to tools used in computing, including word processors, grammar checkers, electronic resources, and the Internet. Specific tools refer to tools used by translators and interpreters, including data capture tools, corpus analysis tools, concordancers, translation memory systems, localization software, webpage translation tools, and translation error checkers. See Figure 4.1.

My definition of translation technology in *A Dictionary of Translation Technology* is from an academic perspective. In the dictionary, translation

DOI: 10.4324/9781003376491-5

Future of Translation Technology   45

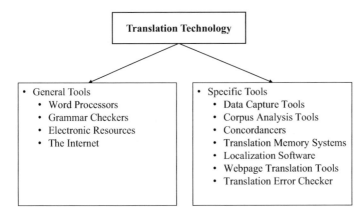

Figure 4.1 Tools in translation technology.

technology is defined as "a branch of translation studies that specializes in the issues and skills related to the computerization of translation" (Chan 2004: 258). After a lapse of 15 years and with the rapid advances in fields related to translation technology, such as computing and terminology, translation technology has now become an umbrella concept that covers the development and use of tools and systems in translating and interpreting. In other words, in addition to tools, systems are given great emphasis. Under this umbrella concept, there are four areas of translation technology: machine translation, computer-aided translation, localization, and speech translation.

Figure 4.2 shows that all the four types of translation technology can be divided into two major subdivisions: machine translation can be subdivided into (1) general applications systems for the translation of common documents and (2) specific-domain systems for the translation of writings in a specific area or subject; computer-aided translation can be subdivided into (1) computer-aided human translation and (2) human-aided computer translation, depending on the degree of human involvement; localization can be subdivided into (1) software localization and (2) webpage localization; and speech translation can be subdivided into (1) two-person conversation and (2) multiperson conference systems. The divisions and subdivisions of translation technology have formed an elaborate structure of technologies that aim to overcome the language and cultural barriers that have separated different language communities since time immemorial.

It is generally acknowledged that TranTech has now grown into a mature technology that is widely used by people at large and by professionals in particular. This then begs the question: What is it for? A simple and honest answer is that it is for replacing human translators or interpreters. Before it can do so, however, it serves as an aid to human translation and human interpreting.

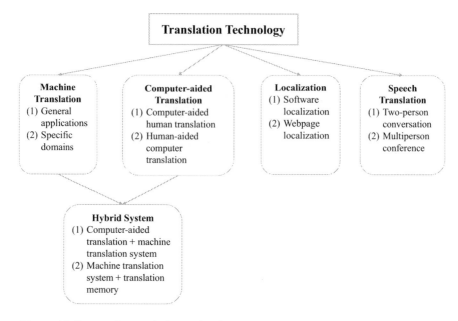

Figure 4.2 Systems in translation technology.

This is evident when we study the development of human translation against different forms of translation technology.

Comparative Study of the Developments of Human Translation and Translation Technology

First, human translation. According to the best information, the earliest form of human translation is interpreting, which began in China in the form of official interpreting in 1122 BCE, totaling 3,145 years ago (Ma 2006). Next is textual translation, which began with the first partial translation of the Bible from Hebrew into Greek in the form of the *Septuagint* (*Hebrew Bible into Greek by Seventy Greek Scholars*), totaling 2,309 years ago. The last is simultaneous interpreting, which began in 1945 with the Nuremberg Trials held in Germany, totaling 78 years (Gaiba 1998).

What about translation technology? In comparison with human translation, translation technology has a shorter history. Machine translation began in 1947, one year after the invention of the computer, and is now 76 years old. Computer-aided translation, which began in 1967 after the ALPAC report, is 56 years old. Localization, which began around 1975, is now 48 years old. Speech translation, which began in 1983, is now 40 years old. Put in chronological order, the seven forms of translation are as shown in Table 4.1.

Table 4.1 Developments of the Seven Types of Translation

Number	Type of Translation	Year of Beginning	Number of Years
1	Interpreting	1122 BCE	3145
2	Textual translation	285 BCE	2308
3	Simultaneous interpreting	1945	78
4	Machine translation	1947	76
5	Computer-aided translation	1967	56
6	Localization	1975	48
7	Speech translation	1983	40

Table 4.2 Division of Work among the Four Types of Translation Technology

Type	Features	Target User
Machine translation	Textual translation Automatic Fast Bilingual/multilingual Information translation	users
Computer-aided translation	Textual translation Interactive Bilingual/multilingual	Professional translators
Localization	Software localization Webpage localization Product globalization	Localizers
Speech translation	Conversational Specific language pair	Tourists Business negotiation

We can observe from the table that, though the human translation period is 45 times longer than that of translation technology, the former has remained much the same throughout its long period of existence, while the latter has brought great and far-reaching changes to the world of translation, resulting in a technological revolution that affects every aspect of translation.

Division of Work in Translation Technology

What is noteworthy is that there is a clear division of work among the four types of translation technology. Machine translation is mainly information (textual) translation for everyone, computer-aided translation is textual translation for professional translators, localization is software and webpage translation for product globalization, and speech translation is for dialogue interpreting for a specific pair of languages.

Several observations can be made from Table 4.2. First, each form of translation technology is serving its own functions in a satisfactory manner. Second, this work division might change in the future due to technological advances. Machine translation, for example, has now been accepted as an effective way of preparing draft translations for professional translators other than its function of information translation for people at large. Third, it is expected that with the proliferation of useful data and automating functions, computer-aided (human) translation will soon move up to the stage of human-aided (machine) translation. Fourth, it is likely that speech translation will be the most popular tool for communication in the future, going beyond dialogue interpreting to more elaborate speech activities, such as business and political negotiations. Fifth, screen enlargement of speech translation systems will make it easier and more effective for users to communicate.

Critical Analysis of the Four Types of Translation Technology

An analysis of the advantages and disadvantages of the four types of translation technology is essential for a better discussion of the future of translation technology.

Machine Translation

As we all know, machine translation (MT) is the automatic production of a target text from one language into another by a translation system. In the area of machine translation, translators and interpreters are users, not players. In other words, they have no role to play in the design and development of MT systems, at least for the time being. Basically, they get what is given.

It is worth noting that machine translation has been a fast-growing area. In 1957, when machine translation came into being, six countries were engaged in the development of translation systems. In 2019, 31 countries are working on it. The huge increase in the number of countries engaged in machine translation and the fast development of systems for different languages and language pairs show that machine translation has advanced by leaps and bounds in the last seven decades. It is therefore important for us to know how machine translation systems work as currently there are hundreds of millions of users around the globe. Google Translate, for example, processes over 100 billion words a day.

In the last seven decades, experts in machine translation, in their effort to search for fully automatic, high-quality translation (FAHQT), have tried more than 20 approaches in system design and construction. Here, we will discuss six important approaches: direct MT, knowledge-based MT, rule-based MT, example-based MT, statistical MT, and neural MT. Have they succeeded in achieving the goal of FAHQT? We might be able to have a better idea of their degree of achievement by studying chronologically the six major approaches and analysing their strengths and weaknesses.

The first attempt started with direct machine translation (DMT) in the early 1970s. Due to factors such as computer capacities and data size, specialists in machine translation relied heavily on the dictionary as a means of text generation. They used dictionaries in translation systems for matching and mapping. There were therefore two types of direct MT: direct matching and direct mapping.

Direct matching is the use of a dictionary to match a source item with a target item to produce a translation, a strategy in machine translation for a specific pair of languages. The quality of this easy-to-design dictionary-based word-by-word machine translation system is closely related to the quality of the dictionaries used for text production. What is undesirable is that this word-by-word translation system is not context-bound, and the most frequently used meaning or primary meaning may not be the optimal meaning for a specific context.

Direct mapping the generation of a translation by carrying out meaning analysis, i.e. mapping, through a ten-step procedure, as shown in Figure 4.3.

Direct mapping systems are easy to design, and they do not require any linguistic theories or parsing principles. The major drawback of this approach is that the target text that comes out from the system follows the syntactical structure of the source language, which is not appropriate.

Another approach that was also used in the 1970s was the knowledge-based machine translation (KBMT). This refers to the kind of machine translation which attempts to apply knowledge engineering techniques to simulate the various knowledges that human translators have for use in the computer. With this kind of system, the computer should be able to disambiguate, to process illogical expressions, and to find implicit meaning from insufficient information. To enable its functioning, two types of knowledge databases have to be built: a common-sense knowledge database and a specialized knowledge database.

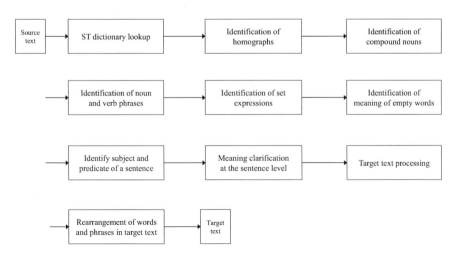

Figure 4.3 Direct mapping procedure.

It is to a certain extent true that a knowledge-based system should be more intelligent than other types of systems. But obviously some issues need to be further addressed. First, as "common knowledge" is not universal, it is hard to build the common-sense knowledge database. Second, "specialized knowledge" may not suit the specific area to be translated, and even if it does, the specialized knowledge database that is built can only be used in a limited way. Third, what is taken to be a typical pattern of behaviour may vary from person to person, and several scenarios have to be worked out for options to be chosen. But a system has no way or context to base its decision on what is the optimal equivalent for the source text/sentence.

The year 1984 witnessed the emergence of two major approaches to machine translation: example-based machine translation (EBMT) and rule-based machine translation (RBMT).

Example-based machine translation is an approach in machine translation that is based on examples (Aramaki et al. 2005, Auerswald 2000: 418–27, Carl and Way 2003, Somers 1999: 113–57). The theoretical basis of this approach was "translating through analogy", which was first suggested by Makoto Nagao in 1984 (Nagao 1984: 173–80). It was thought at that time that advances in computer technology made it possible to gain access to the huge corpora of previously translated analogous examples to allow the matching of bilingual expressions.

Example-based translation goes through two stages: Stage 1 is intralingual matching, and Stage 2 is interlingual matching. In Stage 1, an intralingual matching engine with a database of examples in the source text language is aligned with input sentences. In Stage 2, an interlingual matching engine is used to produce translations of the aligned sentences in a bilingual example database, and translations are produced by the best-match algorithm. See Figure 4.4.

Another approach that emerged during the same period was the rule-based machine translation (RBMT) (Charoenpornsawat et al. 2002, Elming 2006, Proszeky 2005: 207–18, Zhu and Wang 2005). This is generally regarded as a relatively traditional machine translation method which depends on the preparation and maintenance of a large number of rules and lexical information given in general and specialized dictionaries. It is basically a transfer procedure through "word reordering".

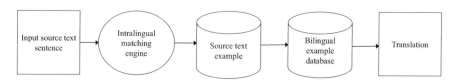

Figure 4.4 Example-based machine translation procedure.

Figure 4.5 Syntactical transfer in RBMT.

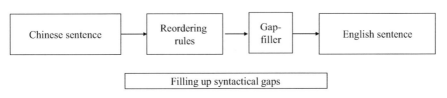

Figure 4.6 Gap-filling in RBMT.

The use of rule-based MT serves two purposes: First, there is a transfer of syntactical structures from the source language and the target language. In translating between Chinese and English, for example, this means converting the syntactic structures of Chinese sentences into the structures of the equivalent English sentences by the use of some reordering rules, which rearrange the words or characters in the source text in the order of the target text. See Figure 4.5.

Second, with the use of a gap-filler, rule-based machine translation bridges the structural gaps that exist during the transfer from one language to another. See Figure 4.6.

Rule-based machine translation, which uses rules to govern text generation, produces "exact translations" of the original as desired by the rule writer, not "probable translations" based on statistics drawn from data. But it can be used together with a probability-based approach, such as statistical machine translation, to produce relatively acceptable translations.

However, there are disadvantages in using the rule-based approach. First, rule writing is time-consuming, labour-intensive, and costly. Second, a clash of rules may occur with the increase in the number of rules. Third, rules cannot be written for long sentences with a large number of strings. Fourth, rules are language dependent or language specific; they are not generalizable to other languages.

In 1991, statistical machine translation (SMT) came into being (Al-Onaizan and Papineni 2006, Knight 2003: 17–19, Koehn et al. 2003, Koehn, et al. 2007, Koehn 2010, Williams et al. 2016). In that year, a group of researchers at IBM TJ Watson Research Center launched the Candide project to reintroduce statistical techniques to machine translation due to the enhanced capacities of the computer. From then on to 2012, statistical machine translation was

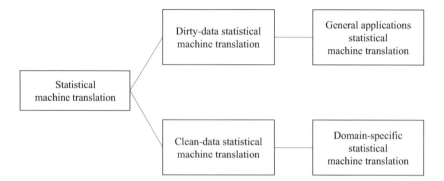

Figure 4.7 Divisions of statistical machine translation.

the main approach in machine translation, until it was gradually replaced by neural machine translation. See Figure 4.7.

What is statistical machine translation? According to Liu Yang and Zhang Min (Liu and Zhang 2015: 201–12), "Statistical machine translation is a machine translation paradigm that generates translations based on a probabilistic model of the translation process, the parameters (translation knowledge) of which are estimated from parallel text." Statistical machine translation uses statistics- or probability-based techniques to perform part of the tasks of machine translation, such as word disambiguation.

Statistical machine translation can be divided into two divisions: dirty-data SMT for text generation in general-purposes systems, such as Google Translate, and clean-data SMT for text generation in domain-specific systems.

Statistical machine translation uses statistics- or probability-based techniques to the machine processing of languages, and these techniques can theoretically be applied to all languages. But at the same time, we should note that translation is about what is contextually accurate, not what is statistically probable.

The latest trend, as we all know, is neural machine translation (NMT), which began in 2013 (Chen and Zhang 2018). This approach has been adopted by major developers of online translation systems, such as Google, Microsoft, and Systran. It is a corpus-based approach that builds up an attention-based encoder-decoder architecture to learn how to translate source sentences into target sentences. With the use of neural machine translation, it is claimed and believed that the quality of translation has improved greatly.

Before we test the quality of machine output of neural machine translation, we should look at the factors that guarantee good machine translation output.

First, data coverage. In the age of Big Data, one of the most important factors affecting the performance of a machine translation system is data coverage, which refers to the amount and type of data stored in the system to cope with the input text.

Second, language kinship. The closer the languages, the better the performance of machine translation.

Third, scope of vocabulary. A large vocabulary avoids the occurrence of "out of vocabulary" (OOV).

Fourth, the well-formedness of the source text. A machine translation system is unable to process a text that is grammatically wrong or syntactically improper.

Fifth, domain specificity. Poor translations will be generated when a domain-specific text is processed by a general-applications system. Likewise, a domain-specific system cannot translate a general text adequately.

Sixth, contextualized translation. It is generally agreed that machine translation does not have the ability to do contextualized translation.

Seventh, length of the sentence. Machine translation is poor at translating long sentences. Shorter sentences are easier to be processed.

Eighth, translation of proper nouns. Machine translation cannot distinguish between common nouns and proper nouns. Semantic translation of proper nouns is obviously unacceptable.

Ninth, translation of abbreviations. The processing of acronyms, for example, is difficult for machine translation.

Tenth, translation of tense. The tenses in different languages are expressed in different ways, and machine translation is not capable of processing tenses adequately.

These ten factors determine to a large extent the quality of machine translation.

Like all other systems, neural machine translation has to handle these ten factors in its own way. A test of two sentences translated by three neural machine translation systems, including Google Translate, Microsoft Translator, and Systran, show that there is room for improvement in Chinese–English translation for all three systems.

Test Sentence 1

> That he quotes Tu Fu to show that the scientist's knowledge of his papers is akin to the writer's is a clear sign that science and art are also kindred: A work of literature is meant for all time. But its ups and downs are known already in the author's heart.

Google Translate output

他引用杜甫來表明科學家對他的論文的了解與作者的相似，這清楚地表明科學和藝術也是相似的：一部文學作品　是有史以來的。但它的起伏是眾所周知的　已經在作者心中了。

Microsoft Translator

他引述 Tu 表示, 科學家對他的論文的知識是類似於作家的是一個明確的跡象, 科學和藝術也是親屬: 文學作品 是所有時間的意思。 但它的起伏是眾所周知的 已經在作者的心臟。

Systran

他引述Tu傅表示，他的资料科学家的知识作家的是如同是一个清楚的标志科学和艺术也亲属：文学工作意味在所有时刻。但是它的盛衰在作者的心脏已经知道。

Human translation

他引用杜甫詩句「文章千古事，得失寸心知」來說明科學家對自己論文之所知與文學家對其作品之所知如同一轍，清楚顯示出科學與藝術性質上並無二致。

Test Sentence 2

想方設法為病人看好病：白衣戰士鍾南山教授

Google Translate

Trying to find a good patient for the patient: Professor Zhong Nanshan, the white soldier

Microsoft Translator

Finding ways to keep a patient optimistic: Professor Bai Zhong Nanshan

Systran

Does everything possible to favour the sickness for the patient: White-coated warrior Professor Zhong Nanshan

Human translation

"Striving My Best to Cure Patients": Professor Zhong Nanshan, A Warrior in White

It is not necessary to make detailed comments on the quality of translations of the two test sentences generated by the three popular neural translation systems. Suffice it to say that much improvement on the systems is needed to produce good translations of the input texts.

The Future of Machine Translation

Based on the preceding discussion, several observations on the future of machine translation can be made. First, can machine translation be used to translate literature? The answer is yes and no. Yes, because machine translation, theoretically, can translate any type of writing, including literature, such as poems and novels. No, because the market may be small, and investing in literary translation is not worth it. Second, can machine translation be a help to professional translators? In the past, machine translation was despised by translators due to its poor output. Nowadays, 56% of professional translators in Europe use popular machine translation systems to prepare draft translations for them to work out acceptable and even publishable translations. The situation in other parts of the world will be similar in the future. Third, it is expected that machine translation will make rapid advances in the future because of (1) advances in artificial intelligence, (2) the increase of big data, including all the e-resources (e.g. e-books), and (3) new developments in neural science. Fourth, machine translation systems will become open systems in the future. All the current machine translation systems are closed systems. They are not user-friendly as they do not allow users to make any adjustments to its database to meet to their needs. In other words, what is wrongly translated will remain so, and users can only repeatedly correct the same errors in the output. Things would improve enormously if machine translation systems could be open systems, allowing users to make lexical, syntactical, and textual changes without affecting the functionality of the system. Fifth, what is the role of artificial intelligence (AI) in machine translation? AI is the science of developing computer systems capable of carrying out human tasks. This refers to the capacity of a machine to replicate the functions and operations of the human brain such as reasoning and learning. Our testing of translation systems shows that machine translation is at the Weak AI stage, and it will be of great help to the translation profession when it advances to other stages (Figure 4.8).

Computer-aided Translation

The future of computer-aided translation is related to its development in the last 56 years, from 1967 to 2023. Central to computer-aided translation is the concept of reusability, which is to use past bilingual documents to prepare data for the translation of current and new documents, as shown in Figure 4.9.

From 1967 to 2019, there have been four periods of development in the field of computer-aided translation, as shown in Figure 4.10.

In the present period of global development, a number of countries in different continents have contributed to the growth of the field, as shown in Figure 4.11. It can be seen from the diagram that the United States in North America is the leader in system development, followed by Japan, Germany, China, the United Kingdom, Canada, and France.

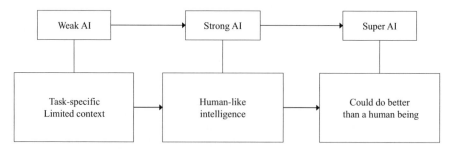

Figure 4.8 Three stages of artificial intelligence.

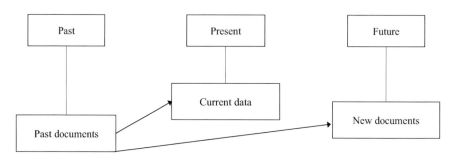

Figure 4.9 Reusability in computer-aided translation.

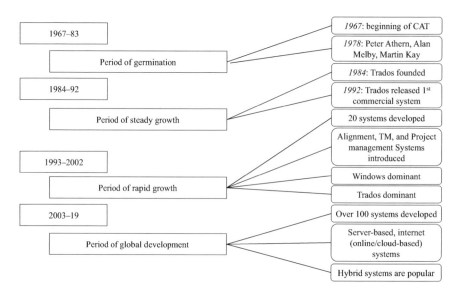

Figure 4.10 Periods of development of computer-aided translation.

Future of Translation Technology 57

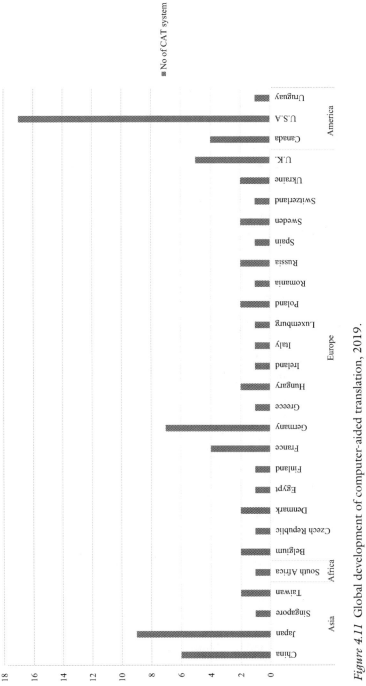

Figure 4.11 Global development of computer-aided translation, 2019.

In the use of systems, cloud-based computer-aided translation systems are fast becoming the major form of Internet-based computer-aided translation systems. Many reasons account for their popularity:

1. Easy accessibility: The system you use is readily accessible anywhere, anytime.
2. Collaborative translation: Files can be shared instantly among members of translation project teams.
3. No software installation needed.
4. Its platform is accessible via a web browser.
5. No system upgrade is needed.
6. Users use the latest and automatic updated system(s) provided by the developer(s).
7. No system incompatibility: All users use the same system provided.
8. No data incompatibility: All users save and use their data on the cloud.
9. Easy to use.
10. Has an automated e-mail system.
11. Real-time progress report is available.

There are two major disadvantages of using cloud-based computer-aided translation systems:

1. Security: Data storage on the Internet is far from safe despite strong assurances from suppliers on Internet security.
2. Subscription: Subscription is long-term, which brings up the concern of affordability for some users.

Future of Computer-aided Translation

Computer-aided translation will continue to be used by increasing numbers of translators. One major reason for its popularity is that, with a very large translation memory database, translation generation can be done by selection, not by translation. This means translation can be done without translating. Terminological consistency in team translation is another advantage, and this is impossible in the case of human translation. Cost reduction as a result of reusing past translations is yet another advantage worth mentioning. Higher volume is achieved for the same reason. And it is much easier to manage translation projects with the use of the project management platform.

At the same time, improvement should be made in translation initialization, which is rather time-consuming. The creation and maintenance of terminology and translation memory databases also need much time and effort. Another concern is about the reusability the stored data, which may be questionable in many cases, especially for ad hoc translators who might be asked to work on a translation project on a specific topic one day but on a different translation project on another day. The speed of text generation, on the other hand, needs further improvement to make computer-aided translation comparable to machine translation.

Localization

Localization is the customizing of software or webpages for a "local" audience, which involves making a product economically, linguistically, technically, culturally, and legally appropriate to the target locale where it will be used and sold. The main purpose of localization is to maximize the understandability and usability of a product so that it can be used in different parts of the world with optimal effectiveness.

Localization can be divided into software localization and webpage localization. Software localization refers to the process of adapting a product or, more specifically software, to a specific locale, i.e. to the language, cultural norms, standards, laws, and requirements of the target market, such as the translation of screen texts and help files. Webpage localization, also known as website localization, refers to the translation of a website originally prepared in a foreign language into another in order to present the contents more effectively to the local population. As aptly put by Anthony Pym (2010: 1), software localization and webpage localization are actually much the same:

> The localization of a website differs from non-hypertext translation with respect to the identification of translatable elements, the tools needed to render them, their non-linearity, the way in which the translation process is prepared and coordinated, and the extent of the changes that may be introduced.

Future of Localization

The future of localization is promising, as the localization of software and webpages is a huge business for the translation industry. The localization of video games, in particular, is becoming increasingly important and profitable. Looking toward the future, a number of areas deserve our examination.

First, it is hoped that localization could be made simpler for translators so that all that needs to be done can be done independently by them, without the help of the technical personnel to deal with the computational aspect of localization and the involvement of the local people to advise on the cultural aspect of localization. This means that the internationalization staging should be better prepared so that computer glitches do not occur while the globalization of the product is made more culturally neutral.

Second is the shift from the English language to strategic languages based on market demands. In recent decades, there have been changes in the languages for localization. In the past, English has been predominantly the language from which localization was made into other major languages, such as French, German, and Spanish. Then there was the emergence of "reverse localization", or localization into English or other major languages. Later still, people started to emphasize "strategic languages", which represented new market areas with a potential for new revenue streams, as opposed to keeping a number of "maintenance languages", such as French, Italian, German, and Spanish (FIGS), whose market had to be maintained and served but with little potential for market growth. More recently, due to the size of its market, the Chinese language has become the number one localization language.

Third is the degree of localization. According to Singh and Pereira (2005), there are five degrees of web localization: "standardized", "semi-localized", "localized", "highly localized", and "culturally customized". The differentiations among these lie in the necessity for translation, which is essential for the "localized" and "highly localized" options. This means that, in the future, when the world becomes more globalized, partial localization rather than full localization will be the norm.

Fourth is the way localization is done. Localization can be done in-house, by localization companies and language-service vendors through outsourcing, by online translation systems such as Google Translate or the online memory-based Google Translator Toolkit, and by non-professional translators through "crowdsourcing". This implies that more words will be translated in a shorter period of time at a lower total cost through the use of global information management systems that are on the market or are tailor-made for companies.

Fifth is making localization a branch of translation studies. At present, localization is generally regarded as an operational procedure in the translation industry, not an academic topic that deserves the attention of translation scholars. It is noted that, during the 11 years from 1995 to 2006, a total of 127 works were published on localization, accounting for only 1.51% of all the literature on translation technology (Chan 2008). This shows that much effort is needed to examine the commercial, computational, linguistic, and theoretical aspects of localization.

Speech Translation

Speech translation, as mentioned at the beginning of this chapter, has been with us for 40 years. That is a relatively short time compared to machine translation or computer-aided translation. Speech translation works between two languages or a single language pair. The source language is the voice-in input, and the translated language is the voice-out output. Voice data or speech signals are essential for speech recognition, speech processing, as well as speech generation. According to a survey, the larger the speech data, the better the performance of a speech translation system. Speech translation systems are usually used in tourism and conversational discourse, such as dialogue interpreting systems. As in the case of machine translation, domain-specific systems are more reliable, e.g. a currency conversion system. It is noted that research has begun on computer-aided interpreting systems to help interpreters in their work.

Future of Translation Technology 61

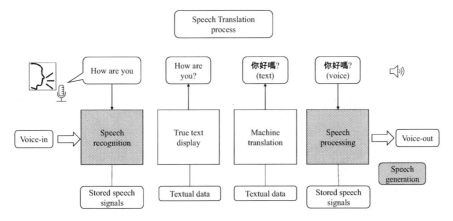

Figure 4.12 Speech processing in a speech translation system.

The process of speech translation is fairly elaborate and complicated. A simple diagram in Figure 4.12 shows how a speech translation system works. When a voice input is made through a microphone, say "How are you?", this is recognized by the speech recognition system based on the stored speech signals. "How are you?" is then displayed on the screen of a device, usually a mobile phone, in the form of textual data. The textual data is then translated by a machine translation system as 你好嗎, which is then processed by a speech processing system with stored speech signals, and the signals are then generated by a synthetic voice in the form of a voice-out. On the surface, the procedure is simple, but the underlying principles that work behind the screen are very complicated.

Future of Speech Translation

Speech translation will grow enormously in the future as speech translation systems are convenient to use and are widely used. The number of users has been on the increase, overtaking other forms of translation technology. The e-speech of Google Translate, for example, has more than 500 million users.

But the number of languages currently available, which is estimated to be around 40, is relatively small, when gauged against the total number of world languages and languages for machine translation, around 150. What is more, the translation that comes out from a speech translation system is mostly unidiomatic, except in the case of phatic language, such as "How are you doing?" Another drawback is that, like machine translation, the speech signals in a speech translation system are not accessible to the user, and there is no way for a user to revise the speech corpus stored in the system. It is hoped that an open system will come along that allows changes in data to be made in order to improve the quality of speech generation.

Conclusion

In revisiting the future of translation technology, conclusive remarks have been made for the four types of translation technology. We have come to the point where we have to look at the future of translation technology from a holistic perspective.

The first observation we have is that computer-aided translation should be taught and learned. It is generally believed that TranTech will be with us in many years to come. As machine translation and speech translation are still black boxes to users, the best way for us to cope with the challenges in the future world of translation is to master the skills of computer-aided translation and use them to the best effect.

There are also professional and entrepreneurial reasons for learning computer-aided translation. From a professional perspective, it should be noted that virtually all translators in technologically advanced countries are computer-aided translators and that most of them are certified or accredited. As rightly pointed out by Timothy Hunt: "Computers will never replace translators, but translators who use computers will replace translators who don't". What is happening in the field of translation technology shows that Hunt's remark may not be far off the mark. In the 1980s, very few people had any ideas about translation technology or computer-aided translation. Nowadays, SDL-Trados alone has more than 200,000 computer-aided translators, and the total number of computer-aided translators in the world is likely to be several times higher than the SDL-Trados translators.

From an entrepreneurial perspective, we should know that translation is in part vocational, in part academic, and in part commercial. In the training of translators, there are courses on translation skills to foster their professionalism, and there are courses on translation theories to enhance their academic knowledge. However, there are very few courses on translation as a business or as an industry. It should be noted that translation in recent decades has increasingly become a field of entrepreneurial humanities as a result of the creation of the project management function in computer-aided translation systems. This means that translation is now a field of humanities which is entrepreneurial in nature. Translation as a commercial activity has to increase productivity to make more profits. It is evident that the impact of computer-aided translation on translation business is far-reaching. Computer-aided translation has made it possible for translation companies to translate faster and better at lower cost.

The second observation we have is that the age of mobile translation will be with us soon. We believe mobile translation will be a dominant trend in the future for the following reasons: (1) The trend of machine translation is from desktop to mobile; (2) with more inventions of digital devices, more mobile translation platforms will be available to us in the future; (3) the large number of mobile phone users creates a huge demand for mobile translation; and (4) the huge pool of bilingual/multilingual talents around the world will be able to use translation apps to make their contribution to mobile translation.

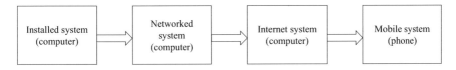

Figure 4.13 Major trends of computer-aided translation.

We can therefore safely conclude that, with the growing popularity of mobile phones, mobile translation systems will be the major trend in the future. This trend is shown in Figure 4.13.

The third observation is that crowdsourcing will become a major trend in translation (Cai 2018: 236–54). Translating texts by crowdsourcing has become increasingly popular, at least in China. Despite the lack of control over quality, this form of translating speeds up the translation of texts manifold. If issues such as quality control and project management could be dealt with in a satisfactory manner, the future of crowdsourcing will be promising.

The fourth observation is the rapid rise of image translation. This is yet another emerging trend of great importance. It is most likely that people in the next generation and beyond will be interested more in images than in texts. Visual translation helps visually handicapped people by means of audio translation. Visual translation can be image-to-text or text-to-image. One important feature of camera translation is its ability to recognize and translate the text in a photo. It can also recognize an object in a photo, produce the name of the object, and translate the name into the target language. With the increase of visual data, great advances will be made in this area.

These four observations are my thoughts on the future of translation technology, in addition to the ten directions mentioned in my book.

References

Al-Onaizan, Yaser and Kishore Papineni (2006) "Distortion Models for Statistical Machine Translation", *Proceedings of the Joint Conference of the International Committee on Computational Linguistics and the Association for Computational Linguistics (COLING/ACL-2006)*, Sydney, Australia.

Aramaki, Eiji, Sadao Kurohashi, Hideki Kashioka, and Naoto Kato (2005) "Probabilistic Model for Example-based Machine Translation", *Proceedings of MT Summit X*, Phuket, Thailand.

Auerswald, Marko (2000) "Example-based Machine Translation with Templates", Wolfgang Wahlster (ed.) *Verbmobil: Foundations of Speech-to-speech Translation*, Berlin: Springer Verlag, 418–27.

Bowker, Lynne (2002) *Computer-aided Translation Technology: A Practical Introduction*, Ottawa: University of Ottawa Press.

Cai, Yixin (2018) "Crowdsourcing translation in contemporary China: Theories and practices", in Chan Sin-wai (ed.) *The Human Factor in Machine Translation*, London and New York: Routledge, 236–54.

Carl, Michael and Andy Way (2003) *Recent Advances in Example-based Machine Translation*, Dordrecht: Kluwer Academic Publishers.

Chan, Sin-wai (2004) *A Dictionary of Translation Technology*, Hong Kong: The Chinese University Press.

Chan, Sin-wai (2008) *A Topical Bibliography of Computer (-aided) Translation*, Hong Kong: The Chinese University Press.

Chan, Sin-wai (2013) "Approaching Localization", in Carmen Millan and Francesca Bartrina (eds.) *Routledge Handbook of Translation Studies*, London and New York: Routledge, 347–62.

Chan, Sin-wai (2017) *The Future of Translation Technology: Towards a World without Babel*, London and New York: Routledge.

Charoenpornsawat, Paisarn, Virach Sornlertlamvanich, and Thatsanee Charoenporn (2002) "Improving Translation Quality of Rule-based Machine Translation", *Proceedings of the Workshop on Machine Translation in Asia (COLING-2002)*, Taipei, Taiwan.

Chen, Jiajun and Zhang Jiajun (eds.) (2018) *Machine Translation: 14th China Workshop, CWMT 2018 Wuyishan, China, October 25–26, 2018 Proceedings*, Singapore: Springer.

Elming, Jakob (2006) "Transformation-based Correction of Rule-based MT", *Proceedings of the 11th Workshop of the European Association for Machine Translation: Machine Translation / Translation Aids—Tools to Increase Quality and to Save Money*, the University of Oslo, Oslo, Norway.

Gaiba, Francesca (1998) *The Origins of Simultaneous Interpretation: The Nuremberg Trial*, Ottawa: University of Ottawa Press.

Knight, Kevin (2003) "Teaching Statistical Machine Translation", *Proceedings of MT Summit IX*, New Orleans, Louisiana, the United States of America, 17–19.

Koehn, Philipp, Franz Josef Och, and Daniel Marcu (2003) "Statistical Phrase-based Translation", Proceedings of the Human Language Technology and North American Chapter of Association of Computational Linguistics 2003 (HLT/NAACL-2003), Edmonton, Alberta, Canada.

Koehn, Philipp, et al. (2007) "Moses: Open Source Toolkit for Statistical Machine Translation", *Proceedings of the 45th Annual Meeting of the ACL on Interactive Poster and Demonstration Sessions*, Association for Computational Linguistics.

Koehn, Philipp (2010) *Statistical Machine Translation*, Cambridge: Cambridge University Press.

Liu, Yang and Zhang Min (2015) "Statistical Machine Translation", in Chan Sin-wai (ed.) *The Routledge Encyclopedia of Translation Technology*, London and New York: Routledge, 201–12.

Ma, Zuyi 馬祖毅 (2006) 《中國翻譯通史》 (*A comprehensive history of translation in China*), Wuhan: Hubei Education Press湖北教育出版社.

Nagao, Makoto (1984) "A Framework of a Mechanical Translation between Japanese and English by Analogy Principle", Alick Elithorn and Ranan Banerji (eds.) *Artificial and Human Intelligence*, Amsterdam: North-Holland Publishing Company, 173–80.

Prószéky, Gábor (2005) "Machine Translation and the Rule-to-rule Hypothesis", in Krisztina Károly and Ágota Fóris (eds.) *New Trends in Translation Studies: In Honour of Kinga Klaudy*, Budapest: Akadémiai Kiadó, 207–18.

Pym, Anthony (2010) "Web Localization", available at www.tinet.cat/~apym/online/translation/2009_website_localization_feb.pdf

Somers, Harold L. (1999) 'Review Article: Example-based Machine Translation', *Machine Translation* 14(2): 113–57.
Williams, Philip, Rico Sennrich, Matt Post, and Philipp Koehn (2016) *Syntax-based Statistical Machine Translation*, San Rafael, California: Morgan and Claypool.
Zhu, Jiang and Wang Haifeng (2005) "The Effect of Adding Rules into the Rule-based MT System', *Proceedings of MT Summit X*, Phuket, Thailand.

5 The Idea of Techno-philosophy and Philosophy-aided Technology, with Social Networking as an Example

Ying Koon Kau

Introduction

The idea of techno-philosophy is still in the process of fermentation and germination. Thanks to the invaluable efforts of Professor Chan Sin-wai, the Techno-Humanities Research Centre was established in Hong Kong, and the research activities of techno-philosophy was kicked off accordingly under the umbrella of techno-humanities. To facilitate this unfolding of techno-philosophy securely as well as sophisticatedly, we must first delineate this very concept in a clear and persuasive way.

The concept of techno-philosophy can be understood in two different ways: as either technology-aided philosophy or philosophy-aided technology.

The first, technology-aided philosophy, is philosophy with the help of technology (especially information technology [IT]). With the rapid progress of IT, philosophers found that philosophical thought experiments (the most well-known example may be the studies of how autonomous vehicles can make appropriate and ethical judgments in real-world road driving, encountering all sorts of possible scenarios) can be conducted more automatically and efficiently by using IT and artificial intelligence (AI). One of these philosophers, Patrick Grim, stated repeatedly that philosophical studies with the help of IT should be propagated (Grim et al. 1998; Grim 2004), while others seconded the notion and hoped that it would help resolve some of the famous moral conundrums (Floridi and Sanders 1999, 2001). And last but not least, philosophers are interested in exploring the basic logical and mathematical principles of IT and AI themselves. We can reasonably expect this technology-aided philosophy to flourish in the coming world. But that is not what I would like to discuss in this chapter.

The second one, philosophy-aided technology, is the concern here, and I will do a preliminary analysis and illustration in this chapter (with the example of social networking). On the one hand, theoretically, we seek to investigate our human condition in the age of technology; we are concerned with the meaning of technology for and the impact of technology on the individual, society, and culture. On the other hand, practically, we have to reflect upon how we should design new technologies that promote the best

DOI: 10.4324/9781003376491-6

values and philosophical ideas for the well-being of mankind. In the example of autonomous vehicles just mentioned, we demand these vehicles (and other similar robots) to comply with the requirements of liability, accountability, and the rule of law. How should we design these vehicles, with the help of so-called machine ethics, to ensure that they "act" ethically? Based on what principles should these machines be manufactured and operated? And, more fundamentally, how must ethics and laws (or the social contract, to use a more theoretical term) be tailored to meet the social and cultural change shaped by this technology? By means of the example of autonomous vehicles, we can see that technology-aided philosophy and philosophy-aided technology actually go hand in hand with each other.

Philosophy-aided Technology

Of course, before the coinage of philosophy-aided technology, related philosophical research had been booming since the advent of modern science and technology. And the source of ideas for all these studies can be traced back to ancient Greece (for example, the analysis of "techne" by Aristotle). Ernst Kapp (1808–96), who was a German American philosopher and geographer, used the term "philosophy of technology" for the first time to denote his philosophical reflection upon modern technology (Kapp 1877). After that, many philosophers and social scientists conducted research in this direction, forming a research area called "humanistic philosophy of technology" (by Carl Mitcham 1994), which means that these kinds of philosophizing originate from humanities and social sciences rather than from the practice of technology itself. Because technology comes from our own human goals and values, these writers sought to investigate the relationship between technology and our human nature, morality, culture, politics, and even metaphysics. "Humanistic philosophy of technology", therefore, takes its point of departure from "the primacy of the humanities over technologies" (Mitcham 1994: 39).

In my opinion, the most famous Chinese philosopher to contribute to this area of "humanistic philosophy of technology" should be Mou Zongsan (牟宗三 1909–95). He was a central figure in contemporary Neo-Confucianism. Since the May Fourth Movement in the beginning of the twentieth century, Chinese intellectuals eventually realized the demands for both science and political reform (democracy), which comprises the motto for the New China in order to have healthier development and modernization. To this end, Mou Zongsan proposed his theory of "Self-Restriction (自我坎陷)", which tries to figure out the relation between Confucian philosophy (and more generally Chinese culture) and science (and democracy). The most significant points in his reflections upon science and technology can be summarized as follows (Mou 1991):

1. In order to learn science and technology from the West, we need not forsake our own Chinese Culture totally, just as the New Culture Movement

advocates thought. Confucianism (as a scholarly discipline and a sophisticated lifestyle) is not against science. Confucianism and science are simply disciplines of different areas of concern. Confucian philosophy can be compatible with science.
2. Quite the contrary, Confucianism requires science and technology to actualize its values:
   i. "Make proper use of resources and enrich the lives of the people (利用厚生)", as mentioned in the ancient text *The Classic of History*[1] (《尚書》), implying that we should utilize our environment in a sensible way to make a better living.
   ii. "The world is for everyone (天下為公)", as mentioned in the ancient text *The Classic of Rites*[2] (《禮記》), implying that we should treat all human beings as equal and work for their well-being.
3. Therefore, according to Mou, Confucian philosophy and science are "coordinative" (對列). On the one hand, they were constructed on different principles and methods and hence are independent of one another. So, although Chinese culture did not produce anything like modern science in its history, we Chinese people can certainly learn from others and develop our own empirical knowledge of the world. And importantly, Chinese culture or Confucianism specifically will not hinder the absorption of science but demand it, as just mentioned. But on the other hand, when science is put to use (as technology), Confucian philosophy should act as a moral system (this is what Confucianism is good at) to critically assess its meaning and deliberately direct its utilization.

Naturally, many philosophers react differently with positive or negative evaluations towards Mou's arguments. Still Mou's ideas are relevant to scholars today, in the area of philosophy-aided technology. Some of them develop, with the help of related Chinese philosophical concepts, into a "Confucian Ethics of Technology" (Wong 2012).

Anyway, if we carefully examine the standpoints of all writers in the domain of "humanistic philosophy of technology", we find out that they can be grouped into three categories.

1. Neutrality Thesis

This group of writers deem technology as a neutral instrument that users can put to good or bad use. So I call their standpoint the "neutrality thesis". Like many Chinese scholars of previous generations, Mou supported the neutrality thesis. He thought that, since natural science is value-neutral, whether science is put to produce weapons (the atomic bomb, for example) or make life better is not something science itself should be responsible for. Technology is therefore a neutral instrument. Our free will (our human ability) should take control and make ethical judgements in utilizing it.

Although a number of scholars held this kind of neutrality thesis (e.g. Pitt 2000), it came under heavy criticism from different parties in the twentieth century, for example from the phenomenology movement (Martin Heidegger being the most prominent figure) and the Frankfurt School (Theodor W. Adorno and Herbert Marcuse). To put it in simplest terms, although science sets out to investigate the empirical world from an objective viewpoint, actual scientific research done from an absolutely unbiased point of view is simply unattainable, thanks to the scrutinization of hermeneutics and the philosophy of science. Nevertheless, technology is quite different from theoretical science in that it aims clearly and frankly to change the world. Technological products by definition have certain functions and aim to alter our lives and/or the world. We make technology, and, in turn, technology remakes us. That is why technology is inarguably value laden and goal oriented. It is not neutral.

2. Optimistic Thesis

Another group of writers thought of technology as an indispensable tool for the progress of human culture and society, and so they are more optimistic towards technology. Karl Marx (1818–83) was the most noteworthy example in this line of thinking. According to his historical materialism and so-called economic determinism, the development of technology (the spinning machine and stream engine were the most valued "means of production" in his age) is one of the indispensable fuels to social progress and human flourishing. This does not mean that technology will not do bad things (especially in the bourgeois mode of technological production), but Marx did believe that ongoing technological development and revolutions could nonetheless lead us to the ideal world of socialism and communism.

3. Not-so-optimistic Thesis

But the reality is that writers with a not-so-optimistic standpoint towards technology outnumbered the two previous groups and represent the mainstream in the "humanistic philosophy of technology". These scholars do not necessarily oppose the use of technology, but they are worried to one extent or another about the effect of technology on our way of being and our culture at large. "As technological revolutions increase their social impact, ethical problems increase" (Moor 2005). Technology creates new dilemmas and ordeals that may be subtle and that we did not face before. That is why we should be alert to technology but not be led by it. The most influential thinker of this not-so-optimistic thesis is Martin Heidegger, who treated technology as modern humans' problematic way of being that we can hardly get rid of. This way of being sees the world as something to be ordered and manipulated in line with projects and plans. This shows a "will to technology" which confines and impoverishes the human experience of reality in far-reaching ways (Heidegger

1977). Technology changes our lifestyle, not always in an agreeable and healthy manner. Moreover, some scholars are even afraid of the "singularity" (proposed by Kurzweil 2006) that AI in the future may break free of the control of human beings and become hard to predict. No matter how possible or impossible the occurrence of such singularity may be, we nevertheless already live in danger of being the devices of our devices and losing our autonomy or authenticity (Heidegger 1977). So a philosophical critique of today's technology becomes very crucial and important. Technology requires philosophy (i.e. philosophy-aided technology) on account of the fact that our modern way of living and our humanity may be in jeopardy.

Next, we will look into the issue of social networking as an example of such a kind of critique in philosophy-aided technology.

Example of Social Networking

A social networking service (SNS, a.k.a. social networking site) is an online platform for people all around the world to share contents and develop relationships and networks. After the emergence of Web 2.0 which emphasised interactivity and user-generated content, some very famous and influential SNSs emerged and have been evolving and prospering since the first decade of the twenty-first century. The most prominent of them are Facebook, Twitter, and Instagram. Because all things on the Internet keep changing and advancing all the time, "social network" (or nowadays more frequently referred to as "social media") is a vague concept which can only be loosely defined and can even be further categorized into different types (Obar and Wildman 2015). In the following discussion, I can only assume Facebook, Twitter, and Instagram are typical examples of SNS in order to introduce briefly some philosophical and ethical issues and controversies aroused. My discussion does not aim at comprehensiveness but only targets a better understanding of what is going on in the academic discipline of philosophy-aided technology.

1. Right to Privacy

We are all accustomed to uploading our personal information, messages, photos, and videos on SNS to share with friends or anyone else who wishes to see them. But from time to time, we find ourselves regretting that we uploaded something we want to retract later or that acquaintances of ours posted something about ourselves which we do not want to be disclosed. Certainly not all of this information is appropriate for publicizing, because either now you find you personally don't want anybody to know or because that information about you causes you more harm than good. This concerns the issue of Internet privacy which has become a pressing human rights issue. Often, it is too late for us to learn that much of the data we upload on the Internet may be impossible to retract or delete securely and permanently. Moreover, usually we are not even aware of the privacy policy of these SNSs: Who is in control of collecting this data, and what is done with it?

Nowadays information is king. Information about SNS users, maybe without our conscious recognition, has become a valuable property of third parties like corporations. That's why such corporations like Facebook have made a fortunes based on the data they collected and sold. These things are already threatening our own privacy and even basic freedom. It is very well-known now that big data—combined with AI and facial recognition software—has the capacity to intrude on people's lives in unprecedented ways, in some cases on a massive scale. In his best-selling book *Homo Deus*, Yuval Noah Harari asks about the long-term consequences of AI and big data: What will happen to society, politics, and daily life when non-conscious but highly intelligent algorithms know us better than we know ourselves (Harari 2017)? The uncomfortable fact is that, when we as SNS users are forced to be transparent, the corporations enjoying profits from our data are not required to be equally transparent. It goes without saying that what they care about is the maximization of profits. "The only hope for social networking sites from a business point of view is for a magic formula to appear in which some method of violating privacy and dignity becomes acceptable" (Lanier 2010). That's why many philosophers raise serious questions and ask us to rethink and reformulate the basic morals and rules of our society in order to protect it in the new circumstances, reasonably ensuring that our interests or rights are not violated and that social harmony could be attained. We "must formulate a new social contract, one that insures everyone the right to fulfil his or her own human potential" (Mason 1986, 11).

2. *Public Sphere is Alienated*

Not only is our private sphere (privacy) being intruded on, but the public sphere is also under attack. SNS absorbs our attention and alienates us in our real lives. Certainly, SNS gives us online or "virtual" shared activities but far from the close intimate friendships that shared daily lives can give. And the latter is what we really need in order to have an enjoyable and happy life. Nevertheless, we don't have any concluding findings about the quality and value of online friendships because the information technology of SNS is still evolving (Sharp 2012). But it is unquestionable that SNS distracts our attention from the needs of those in our immediate physical environment. "When we must compete with Facebook or Twitter for the attention of not only one's dinner companions and family members, but also one's fellow drivers, pedestrians, students, moviegoers, patients and audience members, the integrity of the public sphere comes to look as fragile as that of the private" (Vallor 2021). Electronic devices "de-world" our relationship with people and things, phenomenologically speaking, by disconnecting us from the full contextuality of everyday life. We are endowed with "shorter attention spans and less capacity to engage with critical argument" (Ess 2010: 114) because we are losing situated and embodied engagement with the real world. Naturally, we are now living in states of "alone together", "alienation in connectedness", and therefore moral engagement is limited, and human relations become trivialized (Dreyfus 1999, 2001).

3. Phenomenon of Echo Chambers

Another related issue is the formation of "echo chambers" or "filter bubbles". Due to the advance of big data and AI algorithms, SNS feeds us with advertisements, news, information, propaganda, and even "friends" with very similar interests and viewpoints to ours. Sooner or later, we shall be imprisoned unintentionally (for some people, deliberately) in these "echo chambers" or "filter bubbles" which are information groups or even silos for like-minded net surfers. This can effectively prevent us from exposure to alternative views. Originally, we thought information technology would promote our knowledge and keep us in better touch with the world. But the opposite may be true: The phenomenon of echo chambers can keep us in isolation and insularity. Fake news and ill-founded opinions flood the web, and some worry that such kind of information silos would "brainwash" users to turn to extremism. We should design our IT (SNS) in a new way to shield against this danger in order to promote more objective and comprehensive horizons and to encourage shared interests and experiences of our world citizens. Along these lines, some have strongly suggested that we should postulate new norms and cultivate new virtues of online civic mindedness in order to stop the self-destructive and irrational tyrannies of mob behaviour (Ess 2010). Otherwise, true consensus and solidarity would be very hard to achieve in this technological age.

4. Society, Technology, and the Importance of Philosophizing

The preceding analysis confirms the standpoint of philosophers holding the not-so-optimistic thesis. One of them, Albert Borgmann who was influenced by Heidegger, claims that technology developed in an unhealthy way "will lead to a disconnected, disembodied, and disoriented sort of life. ... It is obviously growing and thickening, suffocating reality and rendering humanity less mindful and intelligent" (Borgmann 1992:108–9). As previously mentioned, we live in the danger of becoming the devices of our devices (Heidegger 1977). Being human, we have our moral obligation not to fall inauthentically and uncritically into the convenience and brainlessness that technology offers. This is why we have to philosophize about the relationship between technology and us.

There is a theory that represents technology as an independent driver of social and cultural change which more or less goes beyond our control. This is known as "technological determinism". But this theory neglects the fact that technology is also the outcome of our social and cultural reality, which every one of us participates in during our daily lives. We can say that technology is socially constructed, as this or that technological artifact shows up as meaningful and useful only from a prior "technological" mood or attitude towards the world, which in turn is heavily shaped by our values and language, society, and culture. Technology makes sense because we already live in the technological age.

A more realistic description would be that society and technology co-construct each other. They are "co-original" and form an intimate and organic whole. Eventually, technology and the changes it brings out are not inevitable but are open for our evaluation and rectification. Therefore, contrary to technological determinism, most philosophers (phenomenologists especially) see society and technology as co-constitutive to each other. They are each other's condition for being what they are. Moreover, both technology and the technological mood discloses the world in a particular manner. That means that every disclosure of the world through technology is at the same time a concealment of other possible disclosures which may also be meaningful or even more meaningful. This is an area we should pay attention to and take more conscious control of. Technology shapes our world. When we design our technology, we are also designing the world we live in and the sort of humans that we are (or will become). Therefore, we can conclude that all the ethical questions raised here in the area of philosophy-aided technology are mostly also ontological questions: What sort of world and what sort of humans (or in phenomenological language, what way of our being) we are becoming, and are we on the right path? To answer all these questions, we have to first examine carefully our attitudes, moods, beliefs, assumptions, practices, identities, relationships, behind our being—with technology, that is—in order to get a clearer and fuller picture of our being-in-the-world in this technological age when information technology is already pervasive and ubiquitous (Heidegger 1977; Dreyfus 1999).

Conclusion: The Road Ahead

To unfold the research projects of techno-philosophy, the area of philosophy-aided technology would be a good starting point. To conclude this paper, I would like to state some urgent items at the top of the to-do list of philosophy-aided technology.

1. Due to the ever evolving nature of IT, philosophy-aided technology is surely a dynamic enterprise which requires continuous reassessment of the changing situation. Multidisciplinary collaboration among philosophers, social scientists, scientists, and technologists is a must, on the one hand, for philosophers to get the most updated knowledge about science and technology and, on the other hand, for scientists and technologists to confront considerations raised by philosophers and social scientists (Moor 2005: 38).
2. We have to rethink and reformulate the basic morals and rules of our society to protect it in the new circumstances, to ensure that the reasonable

interests and rights of each individual and every party are duly respected, and to have a more forward-looking outlook in the technological development in the future. We "must formulate a new social contract, one that insures everyone the right to fulfil his or her own human potential" (Mason 1986: 11). A sound system of social norms is the cornerstone of building a healthy and harmonious world, both online and offline.

3. On the individual and cultural levels, we should postulate and cultivate new virtues of online civic mindedness, without which online "democracies" will continue to be subject to the self-destructive and irrational tyrannies of mob behaviour (Ess 2010). True consensus and solidarity are what we strive for, and we should try to cultivate them by the positive power of IT. Actually, we can revive the mostly forgotten resources of classical virtue ethics to appraise the situation and propose sensible solutions. There are already research outputs in this direction, for example by using Aristotelian ethics (Vallor 2010), Confucian ethics (Wong 2012), and both (Ess 2009).

4. We should have new phenomenological investigations into our technological being-in-the-world. We should ponder again and again, more carefully and comprehensively, what sort of world and what way of being we would like to have. Some suspect that we are increasingly doomed to relate to the world in a disengaged and disoriented manner, but we can undo this if only we remain alert to the problem and expend enough effort to change it. It is not wise or even possible to escape totally from technology. We can, however, choose to think of a really intelligent way for human beings and technology to actualize one another gracefully and have a true co-evolution (to a post-human species perhaps, according to Kurzweil 2006).

Notes

1 *The Classic of History.* James Legge's translation was cited in the Chinese Text Project, which is an online open-access digital library that makes pre-modern Chinese texts available to readers and researchers all around the world (https://ctext.org/).
2 *The Classic of Rites.* James Legge's translation was cited in the Chinese Text Project.

References

Borgmann, Albert (1992) *Crossing the Postmodern Divide*, Chicago: University of Chicago Press.
Dreyfus, Hubert L. (1999) "Anonymity versus Commitment: The Dangers of Education on the Internet", *Ethics and Information Technology* 1(1): 15–20.
Dreyfus, Hubert L. (2001) *On the Internet*, London: Routledge.
Ess, Charles (2009) *Digital Media Ethics*, Massachusetts: Polity Press.
Ess, Charles (2010) "The Embodied Self in a Digital Age: Possibilities, Risks and Prospects for a Pluralistic (democratic/liberal) Future?" *Nordicom Information* 32(2): 105–18.

Floridi, Luciano L. and John W. Sanders (1999) "Entropy as Evil in Information Ethics," *Etica & Politica*, special issue on Computer Ethics, I(2).
Floridi, Luciano L. and John. W. Sanders (2001) "Artificial Evil and the Foundation of Computer Ethics," in *Ethics and Information Technology* 3(1): 55–66.
Grim, Patrick (2004) "Computational Modeling as a Philosophical Methodology," in Luciano Floridi (ed.) *The Blackwell Guide to the Philosophy of Computing and Information*, New Jersey: Blackwell Publishing Ltd.
Grim, Patrick, Gary P. Mar, and Paul St. Denis (1998) *The Philosophical Computer: Exploratory Essays in Philosophical Computer Modeling*, Cambridge, Mass.: MIT Press, 337–349.
Harari, Yuval Noah (2017) *Homo Deus: A Brief History of Tomorrow*, London: Vintage.
Heidegger, Martin (1977) *The Question Concerning Technology and Other Essays*, New York: Harper Torchbooks.
Kapp, Ernst (1877) *Grundlinien einer Philosophie der Technik: Zur Entstehungsgeschichte Der Cultur Aus Neuen Gesichtspunkten*, Braunschweig: Westermann (Kapp 1877 available online); Jeffrey West Kirkwood and Leif Weatherby (eds.), Lauren K. Wolfe (trs.) (2018) *Elements of a Philosophy of Technology: On the Evolutionary History of Culture*, Minneapolis, MN: University of Minnesota Press.
Kurzweil, Ray (2006) *The Singularity Is Near*, New York: Penguin Press.
Lanier, Jaron (2010) *You Are Not a Gadget: A Manifesto*, New York: Knopf.
Mason, Richard O. (1986) "Four Ethical Issues of the Information Age", *MIS Quarterly* 10(1): 5–12.
Mitcham, Carl (1994) *Thinking Through Technology: The Path between Engineering and Philosophy*, Chicago: University of Chicago Press.
Moor, James H. (2005) "Why We Need Better Ethics for Emerging Technologies", *Ethics and Information Technology* 7: 111–19.
Mou, Zongsan 牟宗三 (1991) 《政道與治道》 (*Authority and Governance*), Taipei: Student Bookstore 學生書局.
Obar, Jonathan A. and Steven S. Wildman (2015) "Social Media Definition and the Governance Challenge: An Introduction to the Special Issue", *Telecommunications Policy* 39 (9): 745–50.
Pitt, Joseph C. (2000) *Thinking about Technology: Foundations of the Philosophy of Technology*, New York: Seven Bridges Press.
Sharp, Robert (2012) "The Obstacles against Reaching the Highest Level of Aristotelian Friendship Online", *Ethics and Information Technology* 14(3): 231–39.
Vallor, Shannon (2010) "Social Networking Technology and the Virtues", *Ethics and Information Technology* 12(2): 157–70.
Vallor, Shannon (2021) "Social Networking and Ethics", in Edward N. Zalta (ed.) *The Stanford Encyclopedia of Philosophy* (Fall 2021 Edition), Stanford, CA: Stanford University Press.
Wong, Pak-Hang (2012) "Dao, Harmony and Personhood: Towards a Confucian Ethics of Technology", *Philosophy and Technology* 25(1): 67–86.

6 Corpus-assisted Translation Learning
Attitudes and Perceptions of Novice Translation Students*

Liu Jianwen, Su Yanfang, and Liu Kanglong

Introduction

Corpora have been viewed as empowering translation students by promoting a shift from teacher-centred to student-centred translation pedagogy (Rodríguez-Inés 2009). Among the various types of corpora, parallel corpora allow students to extract translation equivalents, collocations, and bilingual terminologies (Frankenberg-Garcia and Santos 2000, 2015; Bowker and Pearson 2002) and to inductively explore, analyse, and discover language use and translations in different contexts (Bernardini 2016), as well as learn translation strategies from professional translators (Pearson 2003). The affordances of parallel corpora are believed to facilitate students' critical thinking skills and enhance their translation competence (Frérot 2016). While previous studies on corpus-assisted translation teaching focused primarily on senior-level translation students (Heylen & Verplaetse 2015; Liu 2020), little has been done to learn about the affordances and challenges of parallel corpora for students with little or entry-level translation knowledge. While it is easy to acquire the use of a corpus-assisted translation tool within several weeks, the possibility of mastering an additional language and translating it effectively has been contested (de Cespedes 2019). To this end, this study explored novice translation students' perception about using parallel corpus in translation. The findings of this study is expected to facilitate translation educators to better understand how pedagogy-oriented parallel corpora and corpus-assisted translation teaching can be improved.

Literature Review

Corpus-assisted Translation Teaching and Learning

Corpus-based translation studies (CBTS) advocated by Mona Baker (1993) has evolved from the investigation of translation phenomena to corpus-assisted

---

* This research was supported by the University Grants Committee of Hong Kong under the Competitive Research Funding Schemes for the Local Self-financing Degree Sector (UGC/FDS15/H11/17).

DOI: 10.4324/9781003376491-7

translation pedagogy. Recently, the availability of various types of large-scale corpora makes corpus-assisted translation pedagogy more amenable to the needs of translation trainees with varied language abilities. The major advantage of the corpus approach, as agreed by many researchers, lies in the accessibility of large-scale authentic language resources (Reppen 2010). For teachers, these resources are critical in the preparation of teaching materials, as they can decide on what to teach based on the occurrences of language patterns in corpora rather than on their intuition or experience (Li 2013). For students, the descriptive evidence of language patterns offered by corpora is crucial for dealing with their translation problems (Bowker and Pearson 2002) and thus directly contributes to the accuracy and adequacy of language/translation choices and also the enhancement of students' language proficiency (Santos and Frankenberg-Garcia 2007; Zanettin 2002). By examining a large amount of authentic representative texts (Bowker and Pearson 2002: 9), students can develop their own expertise and, through comparative analysis of corpus data, improve their critical thinking and cognitive ability in the long term (Rodríguez-Inés 2009). In addition, corpora which are sampled from text categories determined on a priori basis exist in the form of structured data, which can provide students easy access to quality data and thus assist them in coping with the growing demands of efficiency in the translation markets (Ørsted 2001).

Among the various types of corpora, parallel corpora are considered the most significant for translation teaching and learning (Kübler 2011; Zanettin 2002). Previous research has shown that parallel corpora are convenient and helpful for students to extract terminology and identify translation equivalents (Santos and Frankenberg-Garcia 2007; Zanettin 2002). Many of the mistakes made by learner translators (e.g. miscollocations or misuse of semantic prosodic words) can be addressed with the help of parallel corpora (Ruiz Yepes 2011: 78). Parallel corpora can also offer students opportunities to learn how professional translators utilise different translation strategies in dealing with translation problems (Pearson 2003; Nebot 2008). While earlier researchers mainly focused on the compilation and pedagogical design of parallel corpora for translation teaching purposes (Bernardini 2003; Wang 2004; Barros and Castro 2017), more scholars have recently attempted to obtain empirical evidence to test the effectiveness of using a parallel corpus in translator training. Starting from 2015, there has been a surge in empirical studies to investigate the use of parallel corpora in translation teaching among undergraduate translation majors or MA translation students (Heylen and Verplaetse 2015). More recently, Liu (2020) conducted both intragroup and intergroup experiments on using English–Chinese parallel corpus in translation with senior-year translation majors in China. He confirmed the efficacy of using a parallel corpus for improving wording and terminology in both Chinese–English and English–Chinese translations. The participants also expressed interest and confidence in the parallel corpus and acknowledged its unique values for Chinese–English translation.

Factors Influencing the Adoption of Parallel Corpora in Translation Learning

Existing studies seemed to offer some empirical evidence in support of using parallel corpora in translation teaching. However, we should take heed of the tendency of reporting the advantages instead of challenges in using parallel corpora. All the studies just cited were conducted with somewhat experienced and advanced translation students. Considering that language abilities play an important part in fostering the competence development of translators (de Cespedes 2019), research on corpus application in translation teaching should also include novice translation learners whose language and translation competences are still at the developmental stage. So far, little has been done to understand the affordances and challenges of parallel corpora for this group of translation learners. Another omission in this line of inquiry is the availability of user-friendly corpora suitable for translation teaching purposes, which is reported as the most prominent obstacle hampering the adoption of corpus-assisted pedagogy (Frankenberg-Garcia 2012). Boulton (2012) also found that the difficulty of using corpus tools might further inhibit students' adoption of corpus in their learning. Compared with monolingual and comparable corpora, parallel corpora are less frequently used in translation teaching due to the lack of available parallel texts (Liu 2020). In addition, it is also found that students' personal factors might also influence the adoption of the parallel corpus. For example, Wu, Zhang, and Wei (2019) pointed out that students with low motivation and technological self-efficacy tended not to use technological tools in their translation practices. Difficulties of adapting to an inductive learning style using a corpus can also inhibit students from corpus use in their study (Yoon and Hirvela 2004). Therefore, more studies need to be conducted to understand the possible factors that hinder students' use of a parallel corpus.

Research Questions

The current study aims to explore how novice translation students perceive the use of parallel corpus in translation teaching or practices. Specifically, two research questions guided our study:

1. How can novice translation students benefit from using a parallel corpus?
2. What are the possible factors that impede novice translation students from using a parallel corpus?

Methods

Context and Participants

The participants in this study were 12 sophomores enrolled in a translation course at a Hong Kong university who voluntarily signed up for the corpus-assisted translation training. The students were taking their first translation course at the time of their training. In other words, they had minimal knowledge of translation and were all novice translation students. Before the training, a background survey was administered. Based on the background survey, the native language of all the participants was Cantonese (a Chinese dialect). All of the participants indicated little to no experience using corpora, and half of the participants did not even know what a corpus was.

The corpus-assisted translation training lasted for five weeks. The first three weeks were the training sessions. The instructor introduced the basic concepts related to corpora and the different functions of parallel corpus use in translation. Students also did some related translation exercises with the help of the parallel corpus to practice their skills and consolidate their knowledge. Week 4 was the tutoring session in which students were allowed to freely explore the parallel corpus and were also encouraged to troubleshoot possible issues that arose. In week 5, an in-class translation test was conducted to examine their knowledge of the various functions of the parallel corpus and to understand their corpus-assisted translation behaviours. The test consisted of one English-to-Chinese translation task and one Chinese-to-English translation task. Both tasks were to translate a financial news report containing some financial terms. The students were allowed to use the parallel corpus (i.e. TR Corpus, which will be elaborated in detail in the next section) and the prescribed online dictionaries during the test.

After the training, three participants were selected for follow-up interviews based on purposive sampling (Patton 1989). They reported to have different translation experiences, corpus use experiences, and English proficiency levels. It is believed that a detailed analysis of the perceptions of these participants was conducive to revealing shared perceptions toward parallel corpus use in translation. The focal participant profiles are shown in Table 6.1 (names have been anonymised). Informed consent was obtained with all the participants before the study.

Table 6.1 Focal Participants' Personal Profiles

Name	Major	Translation Experience	Corpus Use Experience	Self-reported English Proficiency Level
Candice	English	Translated some interview transcripts	No	Moderate
Charlotte	Chinese	No	No	Poor
Belle	English	Co-translated one scientific article	Some basic knowledge of BNC	Proficient (IELTS 7)

Figure 6.1 Homepage of the parallel corpus.

Corpus Design

A large-scale parallel corpus was used in this study. The parallel corpus is a web-based corpus with a user-friendly and interactive interface, compiled specifically for translation teaching purposes. The corpus covers a wide array of text types, including news, financial reports, features, company profiles, legal documents, and chairman's statements of listed companies based in Hong Kong and Mainland China. Currently, the corpus has a total of around 80 million English words and 172 million Chinese characters, which are annotated for the part-of-speech and aligned at the sentence level to facilitate bilingual concordance and co-occurrence. Modelling after COCA (Corpus of Contemporary American English), TR Corpus also has three primary search functions and one upload function, as shown in the homepage (see Figure 6.1)

Data Collection and Analysis

In order to reveal a holistic picture of parallel corpus use in translation teaching from the perspective of students, a mixed methods approach was adopted to collect multiple types of data.

The data collection and analysis started with an integrative analysis of the survey results and students' search histories on the parallel corpus to explore some key issues underlying corpus-assisted translation teaching. Students' search histories on TR Corpus in completing the translation tests were collected to complement the survey findings. It is believed that observation of students' actual behaviours of using the parallel corpus can help us know when and why the students sought help from the corpus. The search histories were coded based on the types of search they conducted, i.e. words, multiple-word expressions, sentence structures, proper nouns, and terminologies. After the training, the students were asked to finish a 7-point Likert questionnaire survey to examine the perceived usefulness and difficulties of using parallel corpus in translation. The survey was adapted from the survey of Yoon and Hirvela (2004) regarding students' attitudes towards corpus use in L2 writing and from Liu (2020)'s survey about students' assessment of corpus use in translation. Two experienced researchers then reviewed the adapted survey to ensure validity. Descriptive statistics of the mean scores of students' responses were calculated to understand students' perceptions about the affordances and challenges of parallel corpus use in translation.

Based on the initial exploration, subsequent interviews and longitudinal follow-up interviews were conducted to provide a close-up and in-depth understanding of the possible factors that might influence the adoption of the parallel corpus in translation (Miles, Huberman, and Saldaña 2014). Two rounds of semi-structured interviews with three focal students, each round lasting for about 50 minutes, were conducted immediately after the corpus training to supplement and explain the survey results and to delve deeper into students' perceptions about corpus use in translation (Seidman 2006). The interview questions enquired into students' prior experiences, their English proficiency, their translation learning and corpus use, and their perceptions about using parallel corpus in translation. The interviews were conducted via online meetings using Cantonese, Mandarin Chinese, and English, as preferred by the participants Candice, Charlotte, and Belle, respectively.

Then half a year after they finished the corpus training, two focal participants—i.e. Candice and Charlotte, one who kept using and the other who gave up using the parallel corpus—were invited to participate in a follow-up interview which lasted for about 30 minutes for each person. The follow-up interview can help us understand the translation students' perceptions about parallel corpus from a longitudinal perspective. All the interviews were audio-recorded with the participant's consent and were transcribed verbatim. Then the research team scrutinised the interview transcripts and followed the iterative process of initial open coding, focused coding, and categorization until the categories reached saturation.

Finally, the different types of data were compared to identify possible items of discrepancy or corroboration to further explore the possible reasons and answer the two research questions.

Findings

Affordances of the Parallel Corpus

Parallel Corpus as a Reference Resource

Table 6.2 shows students' perceived usefulness and advantages of a parallel corpus. Overall, students reported positive evaluations of the parallel corpus such as solving translation problems at the word, terminological, sentence,

Table 6.2 Usefulness of Parallel Corpus

Usefulness	Number of Agree Responses	Number of Neutral Responses	Number of Disagree Responses	Mean	SD
Meaning of English vocabulary	10	2	0	5.83	0.99
Meaning of Chinese vocabulary	11	1	0	6	0.91
Usage of English vocabulary	10	2	0	5.92	1.04
Usage of Chinese vocabulary	11	1	0	5.83	0.90
Usage of English Phrases	10	2	0	5.92	1.04
Usage of Chinese Phrases	11	1	0	5.83	0.80
Solving terminological problems	10	2	0	5.83	0.99
Solving translation problems at the sentence level	10	2	0	5.83	1.07
Maintaining translation style	10	2	0	5.83	1.07
English decoding skills	11	1	0	5.67	1.18
Chinese decoding skills	10	2	0	5.59	1.32
English encoding skills	11	1	0	5.75	1.01
Chinese encoding skills	9	2	1	5.42	1.44
Confidence in Chinese–English translation	8	4	0	5.50	1.12
Confidence in English–Chinese translation	10	2	0	5.67	0.94
Sense of professionalism	11	1	0	5.58	1.11
Would use the corpus in the future	11	1	0	5.67	0.85

Table 6.3 Search Histories on TR Corpus

Items	E-C	C-E	Sum
Words	96	31	127
Multiple-word expressions	169	181	350
Sentence structures	8	95	104
Proper nouns	51	46	97
Terminologies	115	80	195
Sum	439	445	884

and stylistic levels. The search histories of students, as shown in Table 6.3, confirmed most of the survey results. Each student searched over 70 times on average, indicating the affordances of the parallel corpus as a reference resource in translation. Table 6.3 also shows a prominent number of search records for multiple-word expressions and terminologies in both translation directions. It seemed that students relied on the parallel corpus for solving sentence-level problems in Chinese–English translation, indicating their lack of confidence in using English structures to translate Chinese. In comparison, students exhibited a higher frequency of single words in their search records in English–Chinese translation.

Interviews corroborated with the survey results and search histories. Candice mentioned that the parallel corpus was helpful in translating the terminology of specialized texts. Belle and Charlotte also agreed that the parallel corpus was very useful in searching technical terms. For example, Belle searched for 進出口總值 (*jinchukou zongzhi*) ("the total value of imports and exports") in the parallel corpus.

> I searched for that one because I felt it's a technical term. So I couldn't just use my own translation; I couldn't just use what I think it should be. It's not like some random vocabulary, and I think they must have a proper name for that.

Besides finding translation equivalents, the students also reported that the corpus helped them understand the source texts and enhanced accuracy in the Chinese–English translation. For example, in translating 今年前兩個月 (*jin'nian qian liangge yue*), both Belle and Candice were confused about whether the third character "前" (*qian*) means "before" or "first". By retrieving similar lexical bundles in the parallel corpus, they understood that the phrase should be translated into "in the first two months of this year". Both Candice and Charlotte regarded the parallel corpus as an effective reference tool for solving word-level translation problems.

Aligned with the survey results and search histories, students reported that the sentence structures from the parallel corpus are especially beneficial for Chinese–English translation. For Belle, this was what she valued most in using the parallel corpus. As she mentioned in the interview:

> I tried to look at the structure, for example, whether a word should be put at the beginning or at the end of a sentence. I tried to find out which sentence structure is more frequently used in the translation and which sentence order is more common. And then I would use my version based on those (structures).

During the translation test, she searched the parallel corpus to find an appropriate structure for translating …呈現出四大特點 (*chengxian chu si da tedian*) ("are characterized by the following four features"). Even though the corpus occurrences were not exactly the same as the source text, she could find and analyse some similar sentences to produce an appropriate translation, as previously stated.

At the textual level, all three participants noted that the corpus was vital in ensuring the stylistic appropriateness of translation in specific contexts or genres. For example, when translating "this year' used in a formal news report, Charlotte searched the corpus and found that "this year" could be translated into "本年" (*ben nian*) which is a more formal expression than "這一年" (*zhe yinian*).

One area that attracts our attention is how the students developed critical thinking and analytical skills through corpus-assisted translation teaching. Both Candice and Charlotte mentioned that they pieced together some words found in the corpus to translate a complete sentence. Their translation behaviours indicate an overreliance on the word/phrase equivalents provided by the corpus. Besides, the participants always accepted the results offered by the corpus, because they believed in the reliability of the corpus data. They seldom use the "Compare" function or the "Collocates" function since the Basic Search function already satisfied their queries for translation equivalents (Belle and Candice). Such behaviours indicated that novice translation students might lack the skills to critically analyse the corpus data.

Parallel Corpus as a Learning Aid

Besides using the parallel corpus as a reference tool, the interviewees also mentioned using the corpus as a learning aid in translation. For example, Candice stated that she paid attention to the translation strategies employed by professional translators, such as omissions or part-of-speech transitions in English-Chinese and Chinese-English translations. Belle also repeatedly mentioned that the parallel corpus was a helpful learning aid for L2 learners. She explained that the most valuable affordance of the parallel corpus was not for literal or direct translation but for helping L2 learners to understand how translation was done "in a common way". Corpus examples helped her to know the possible translation norms or taboos.

The corpus also serves as a good learning aid in enhancing students' second language skills. As the survey indicated, most students found the corpus useful in improving their language encoding skills. The focal participants explained how the parallel corpus could be used to improve their writing. For example, Charlotte mentioned that she enlarged her vocabulary and increased the variety and accuracy of word choices in writing. She also used the parallel corpus to search for parallel texts in her English writing:

> I would search for the keywords of a certain topic on the corpus to get some translation examples. The URL links that come together with each example can further direct me to the source websites. In this way, I can get many articles or websites that are useful for my writing.

Belle, likewise, used the parallel corpus to ensure that her writing style was appropriate. Furthermore, the three participants all reported enhanced search skills with increasing use of the parallel corpus. Where Belle initially would just "copy the entire phrase" from the corpus occurrences, she quickly learnt that this approach was not adequate and that adaptation was needed ("to trim it down") to ensure accuracy. In addition, Belle also developed a "double-check" tactic by conducting a further search on the corpus. Specifically, she would search the corpus for the translation candidates provided by the corpus to verify whether such translations are accurate. Such a method of verifying the translation-of-a-translation shows her creativity in using the corpus.

Affective Support

Importantly, the parallel corpus also offered affective support by enhancing learners' confidence in translation and their sense of professionalism (see Table 6.2). It is worth noting that students feel more confident in English–Chinese than in Chinese–English translation. This might be accounted for by the fact that all the participants are native Chinese speakers, so it is easier for them to translate into rather than out of their native language (Liu 2020). As far as confidence is concerned, the corpus can further help translation students with the decoding of English source texts and the encoding of Chinese translations.

In the interview, the translation students all expressed anxiety due to their lack of translation experience, especially when they needed to translate specialized texts that they were unfamiliar with:

> I had no confidence in my own translation. Even when I produced a translation version, I would feel that it might not be as appropriate in wording or style (Candice).

However, Candice noted that she felt more confident with her translation quality after the corpus-assisted translation training because she "has learned how to use the corpus tool, and therefore could make use of more resources than other students [who did not attend the training]". She also noted that, by checking translations in the corpus, she became more confident in the stylistic appropriateness of her translation. With similar sentiment, Charlotte stated that her confidence increased after training which was unavailable to her peer students. Likewise, Belle noted that the parallel corpus offered a kind of "reassurance", since the translation examples provided by the corpus are more comprehensive than the definitions of dictionaries. As the corpus occurrences are displayed in context, they have a clear advantage over dictionaries in which the words are only semantically defined and rarely used in context. By examining the translation examples in the corpus, the students have reported improved confidence in translation.

Affective support also came from the reliability and quality of the corpus data. Both Charlotte and Candice expressed their trust in the parallel corpus data which is more straightforward and cost-effective than web searching or machine translation. Although the corpus might not contain all the technical terms that students wanted to search, the amount of quality data in the form of pre-existing translations by experienced translators can serve as professional references for students to work on their own translation (Pearson 2003).

Factors Obstructing the Adoption of the Parallel Corpus

Although 11 students expressed their intention to use the corpus in the long run in the survey, only four students were found to keep using the parallel corpus six months after the training based on their search histories, indicating that some obstructive factors still exist despite the affordances of the parallel corpus.

Ease of Use

The ease of using a parallel corpus is an important factor affecting students' adoption of it. As shown in the survey results (see Table 6.4), one-third of the participants reported problems in time and effort spent on analysing the data and performing search techniques. Four students also reported difficulties in identifying the proper search techniques. The focal participants further elaborated on some "annoying" characteristics of the parallel corpus that discouraged them from using it for translation. In the parallel corpus, the search keywords can be highlighted while the translations cannot be highlighted because a word or language expression can be translated in different ways. Students thus need to analyse the translations against the source text to find out how a word is translated. However, students frequently mentioned that the translation of the keywords was not highlighted in the result, which added to their difficulty of analysis, especially when the sentence was very long (noted by Candice) or when they were unfamiliar with the context (Belle). In the follow-up interview six months after the training, Charlotte mentioned that online dictionaries or the search engines were more convenient in this regard as she sometimes only wanted a translation of a word or term.

Table 6.4 Challenges in Using the Parallel Corpus

Difficulties/Problems	Number of Agree Responses	Number of Neutral Responses	Number of Disagree Responses	Mean	SD
Time and effort spent on analysing data	4	2	6	3.92	1.66
Unfamiliar vocabulary in corpus output	5	3	4	4	1.63
Too many sentences in corpus output	4	4	4	4.08	1.61
Limited number of sentences in corpus output	4	4	4	4	1.53
Performing search techniques	4	2	6	3.75	1.83

Language Proficiency Level

In evaluating the usefulness of the parallel corpus, five students reported unfamiliar vocabulary in the corpus output as one of the difficulties. For novice translation students, their low language proficiency level is clearly an obstructive factor for the adoption of the parallel corpus.

Charlotte, who gave up using the parallel corpus after the training, reported having more difficulties in using the corpus in Chinese–English translation than in the other direction. Her relatively poor English proficiency, together with a better command of Chinese (as a Chinese major), might explain her experience. Even though the corpus offered her translation examples in various contexts, she still found it challenging to reorganize the English words in proper sentence structures in Chinese–English translation. In contrast, she had performed with better confidence in English–Chinese translation with the aid of the corpus. Charlotte also reported her disappointment: "I could not understand the English text, and I could not find out the translation equivalents". This shows that parallel corpus might be more useful in L2–L1 translation for novice translators as they cannot write or translate effectively in English. Such a finding is contrary to Liu (2020) who found that translation majors found that parallel corpus is more useful in Chinese–English translation than in the other direction.

Candice mentioned in the follow-up interview that, when she needed to translate colloquial Cantonese words into English, she must firstly do an intralingual translation to translate the words into formal Chinese synonyms before searching for the English equivalents in the parallel corpus. However, as a local Hongkonger with a Cantonese background, her low proficiency in written Chinese also limited her use of the corpus.

Practical Needs

Whether students needed to work with a specialized translation was another factor that influenced their continued use of the parallel corpus. Candice and Charlotte reported that the corpus which contains practical translations was not very useful for literary translation, which was often the focus of translation courses. This indicates that pedagogy-oriented corpus design should take the specific needs of translation students into consideration. Although the corpus genres might be an important factor hampering the utilization of corpus use by students, it also brought to our attention the clear gap between a translation curriculum which places too much emphasis on literary translation and the translation industry in which specialized translation accounts for more than 90% of all the translation work (Chan 2015: 44).

This point is best reflected in the case of Candice who first complained that the corpus was not very useful for her schoolwork but who noted that she found it particularly helpful when she had to work with related translations during her internship. In the follow-up interview six months after the training,

Candice reported that part of her internship was to translate between English and simplified Chinese; thus her "needs" for high-quality simplified Chinese–English parallel texts greatly increased, and she was happy to still have access to the parallel corpus. As a result, her perceived usefulness of the corpus also improved. She commented that the Chinese translations provided by the parallel corpus were of high quality and much better than the fragmented or unnatural "Europeanized" translations on the Internet. Contrary to Candice, Charlotte mentioned in the follow-up interview that she gave up the parallel corpus after the training because there was no genuine need for her to work with either Chinese–English or English–Chinese translations.

Discussion

The findings reveal that a parallel corpus could empower novice translation students from different perspectives. Both the survey findings and the focal participant interviews confirmed the benefits of a parallel corpus for translation. For novice translation students, a parallel corpus was most useful as a reference tool at the word, terminological, sentence, and stylistic levels, suggesting the efficacy of the parallel corpus in improving students' translation quality. This is consistent with the findings of previous studies conducted with more experienced or proficient translation students that the parallel corpus serves most effectively as a reference tool (e.g. Heylen and Verplaetse 2015; Liu 2020). In addition, students also used the parallel corpus as a learning aid in translation and language learning, which demonstrated that novice translation students can explore the usefulness of the parallel corpus by themselves, regardless of their language proficiency level. It also indicates the potential of using a parallel corpus to improve students' language skills, which have always been a central component in translation competence development (Li 2013). Students also reported that the parallel corpus offered them affective support, indicating corpus use could help reduce anxiety even though students might have little translation experience. Such affordances of the parallel corpus align with the suggestion that using technological tools could improve students' self-efficacy in translation training practices (Wu, Zhang, and Wei 2019) and reduce students' anxiety by fostering a sense of achievement (Yan and Wang 2012), both of which are important indicators of student performances (Lian, Chai, Zheng, and Liang 2021; Liu 2022). Our findings are largely in line with the findings of Kübler, Mestivier, and Pecman (2018), who found that a corpus enables students to produce better-quality translations despite some technological and pedagogical limitations. In sum, the usefulness of the corpus was well acknowledged despite the varied language abilities of the participants. Belle, a relatively proficient English learner, focused on using the corpus to improve her translation competence or writing skills, while, for Candice and Charlotte, corpus served as a helpful reference resource for translation. This shows that a parallel corpus with its diverse functions can cater to a spectrum of students' needs.

Despite the perceived usefulness of the parallel corpus, some factors are also found to negatively influence students' adoption of the corpus in their translation learning. One major factor is the ease of corpus use. Compared with Boulton's (2012) findings, students in this study reported relatively moderate difficulties in using the parallel corpus, which might be attributed to the user-friendly design of the parallel corpus used in this study. On the other hand, the time and efforts, as well as search strategies needed to effectively use the parallel corpus might lead to technology hesitancy (Wu et al. 2019). Second, students' personal factors also influence their adoption of the parallel corpus. For novice translation students, their relatively low language proficiency clearly affects their analysis and understanding of the corpus data. In fact, previous studies have revealed that L2 language proficiency plays a key role in making effective translation decisions and that, as a result, it should be treated as a prerequisite in translator training (Prior, Whinney, and Kroll 2007; Wu et al. 2019). Furthermore, students' actual translation needs to complete real-world tasks, which can also be a deciding factor for their adoption of the parallel corpus in the long run.

The findings of the study have important implications for corpus-assisted translation training and parallel corpus design. Despite students' positive evaluations of parallel corpus use, interview data also indicated that novice translation students still regarded it as more of a reference tool than a learning aid. Such perceptions might be accounted for by the learners' inadequate language and analytical skills. Students also reported overreliance on corpus results, which were often directly used without critical analysis. This corroborates previous studies' findings that translation students tend to resort to external resources more frequently and process the information more superficially than professional translators (Whyatt 2012). All these problems call for a more comprehensive course design when using a parallel corpus for translation teaching. Since language proficiency is a crucial component of translation competence, translation teachers should strike a balance in their teaching of language skills, translation strategies, critical thinking and analytical skills, and digital literacy (de Cespedes 2019; Li 2002). In teaching novice translation students, teachers can place a special focus on improving students' language skills besides developing other skills and competence. Secondly, although the corpus functions are relatively easy to acquire, performing effective search techniques and analysing corpus results to make proper translation decisions might still be difficult for novice students (Stewart 2009). To help students use the corpus more effectively, teachers can guide students to do a variety of exercises aiming to solve different translation problems (Barros and Castro 2017). Most importantly, teachers can explicitly teach the effective searching techniques in order to get the desired parallel texts (Stewart 2009). Thirdly, collaborative learning, which has proved useful in corpus-based training (Ma, Tang, and Lin 2021) and language learning (Chen and Du 2022), might also be helpful for students who are unfamiliar with corpus methods in translation or at the low language proficiency level. In this study, two participants discovered that the parallel corpus was useful for improving their second language writing skills. This repurposing of the parallel corpus through collaboration has extended its function for empowering novice translators.

Conclusion

This study investigated novice translation students' perceptions and the possible factors that inhibit them from using the parallel corpus in translation in the Hong Kong context. Overall, novice translation students acknowledged the merits of parallel corpus use in the translation classroom. The corpus empowered student translators by serving as a reference tool and learning aid and by providing necessary affective support. However, the ease of corpus use and students' practical needs for translation or English writing are factors that influence their adoption of the parallel corpus in the long run.

It should be acknowledged that this study focused on only a small sample group. The findings can be further tested with a larger group of participants. In future studies, the textual analysis of students' translations can also be undertaken to provide additional evidence and to triangulate data. It is hoped that our study can provide educators and researchers with a better understanding of the role of the parallel corpus for novice translation students. To a large extent, translation teachers need to tailor-make appropriate and effective corpus-assisted teaching activities and materials based on students' needs and levels to ensure that the parallel corpus can make some meaningful impact on the translation classroom.

References

Barros, Elsa Huertas and Miriam Buendia Castro (2017) "Optimizing Resourcing Skills to Develop Phraseological Competence in Legal Translation: Tasks and Approaches", *International Journal of Legal Discourse* 2(2): 265–90, doi: 10.1515/ijld-2017-0015.

Bernardini, Silvia (2003) "Designing a Corpus for Translation and Language Teaching: The CEXI Experience", *TESOL Quarterly* 37(3): 528–37, doi: 10.2307/3588403.

Bernardini, Silvia (2016) "Discovery Learning in the Language-for-translation Classroom: Corpora as Learning Aids", *Cadernos de Tradução* 36: 14–35, doi: 10.5007/2175-7968.2016v36nesp1p14.

Bowker, Lynn and Jennifer Pearson (2002) *Working with Specialized Language: A Practical Guide to Using Corpora*, London and New York: Routledge.

Chan, Sin-wai (ed.) (2015) *Routledge Encyclopedia of Translation Technology*, London and New York: Routledge.

Chen, Chen and Xiangyu Du (2022) "Teaching and Learning Chinese as a Foreign Language through Intercultural Online Collaborative Projects", *The Asia-Pacific Education Researcher* 31(2): 123–35.

de Cespedes, Begona Rodriguez (2019) "Translator Education at a Crossroads: The Impact of Automation", *Lebende Sprachen* 64(1): 103–21, doi: 10.1515/les-2019-0005.

Frankenberg-Garcia, Ana (2012) "Raising Teachers' Awareness of Corpora", *Language Teaching* 45(4): 475–89. doi: 10.1017/S0261444810000480.

Frankenberg-Garcia, Ana (2015) "Training Translators to Use Corpora Hands-on: Challenges and Reactions by a Group of Thirteen Students at a UK University", *Corpora* 10(3): 351–80. doi: 10.3366/cor.2015.0081.

Frankenberg-Garcia, Ana and Diana Santos (2000) "Introducing COMPARA, the Portuguese English parallel corpus", in *Proceedings of CULT*, available at www.linguateca.pt/documentos/Frankenberg-GarciaSantos2000.pdf

Frérot, Cecile (2016) "Corpora and Corpus Technology for Translation Purposes in Professional and Academic Environments: Major Achievements and New Perspectives", *Cadernos de Tradução* 36: 36–61, doi: 10.5007/2175-7968.2016v36nesp1p36.

Heylen, Kris and Heidi Verplaetse (2015) "Parallel Corpora for Medical Translation Training: An Analysis of Impact on Student Performance", in International Conference on Corpus Use and Learning to Translate, May 27-29, Sant Vicent del Raspeig, Spain.

Kübler, Natalie (2011) "Working with Different Corpora in Translation Teaching", in Anna Frankenberg-Garcia, Lynne Flowerdew, and Guy Aston (eds.) *New Trends in Corpora and Language Learning*, London: Bloomsbury, 62–80.

Kübler, Natalie, Alexandra Mestivier, and Mojca Pecman (2018) "Teaching Specialised Translation through Corpus Linguistics: Translation Quality Assessment and Methodology Evaluation and Enhancement by Experimental Approach", *Meta: Translators' Journal* 63(3): 807–25.

Li, Defeng (2002) "Translator Training: What Translation Students Have to Say", *Meta: Translators' Journal* 47(4): 513–31, doi: 10.7202/008034ar.

Li, Defeng (2013) "Teaching Business Translation: A Task-based Approach", *The Interpreter and Translator Trainer* 7(1): 1–26, doi: 10.1080/13556509.2013.798841.

Lian, Jinging, Ching Sing Chai, Chunping Zheng, and Jyh-Chong Liang (2021) "Modelling the Relationship between Chinese University Students' Authentic Language Learning and their English Self-efficacy during the COVID-19 Pandemic", *The Asia-Pacific Education Researcher* 30(3): 217–28.

Liu, Kanglong (2020) *Corpus-Assisted Translation Teaching: Issues and Challenges*, Singapore: Springer.

Liu, Meihua (2022) "Foreign Language Classroom Anxiety, Gender, Discipline, and English Test Performance: A Cross-lagged Regression Study", *The Asia-Pacific Education Researcher* 31(3): 205–15.

Ma, Qing, Tang Jinlan, and Lin Shanru (2021) "The Development of Corpus-based Language Pedagogy for TESOL Teachers: A Two-step Training Approach Facilitated by Online Collaboration", *Computer Assisted Language Learning*, advance online publication, doi: 10.1080/09588221.2021.1895225.

Miles, Mathew B., A. Michael Huberman, and Johnny Saldaña (2018) *Qualitative Data Analysis: A Methods Sourcebook*, California: Sage Publications.

Nebot, Esther Monzo (2008) "Corpus-based Activities in Legal Translator Training", *The Interpreter and Translator Trainer* 2(2): 221–52, doi: 10.1080/1750399X.2008.10798775.

Ørsted, Jeannette (2001) "Quality and Efficiency: Incompatible Elements in Translation Practice?" *Meta: Translators' Journal* 46(2): 438–47.

Patton, Michael Quinn (1989) *Qualitative Evaluation Methods* Los Angeles, California: Sage Publications.

Pearson, Jennifer (2003) "Using Parallel Texts in the Translator Training Environment", in Federico Zanettin, Silvia Bernardini, and Dominic Stewart (eds.) *Corpora in Translator Education*, Northampton, Manchester: St. Jerome Publishing, 15–24.

Prior, Anat, Brian Whinney, and Judith F. Kroll (2007) "Translation Norms for English and Spanish: The Role of Lexical Variables, Word Class, and L2 Proficiency in Negotiating Translation Ambiguity", *Behavior Research Methods* 39(4): 1029–38, doi: 10.3758/BF03193001.

Reppen, Randi (2010) *Using Corpora in the Language Classroom*, Cambridge and New York: Cambridge University Press.

Rodríguez-Inés, Patricia (2009) "Evaluating the Process and Not Just the Product When Using Corpora in Translator Education", in Allison Beeby, Patricia Rodríguez-Inés, and Pilar Sánchez-Gijón (eds.) *Corpus Use and Translating: Corpus Use for Learning to Translate and Learning Corpus Use to Translate* (pp. 129–149). Amsterdam: John Benjamins Publishing, 129–49.

Ruiz Yepes, Guadalupe (2011) "Parallel Corpora in Translator Education", *Electronic Journal of Didactics of Translation and Interpretation* 7: 65–80.

Santos, Diana and Ana Frankenberg-Garcia (2007) "The Corpus, its Users and their Needs: A User-oriented Evaluation of COMPARA", *International Journal of Corpus Linguistics* 12(3): 335–74, doi: 10.1075/ijcl.12.3.03san.

Seidman, I. (2006) *Interviewing As Qualitative Research: A Guide for Researchers in Education and The Social Sciences*. New York/London: Teachers College Press.

Stewart, Dominic (2009) "Translating Semantic Prosody", in Allison Beeby, Patricia Rodríguez-Inés, and Pilar Sánchez-Gijón (eds.) *Corpus Use and Translating: Corpus Use for Learning to Translate and Learning Corpus Use to Translate*, Amsterdam: John Benjamins Publishing, 29–46.

Wang, Kefei (2004) "The Use of Parallel Corpora in Translator Training", *Media in Foreign Language Instruction* 6: 27–32.

Whyatt, Bogusława (2012) *Translation as a Human Skill: From Predisposition to Expertise*, Poznań: Wydawnictwo Naukowe UAM.

Wu, Di, Lawrence Jun Zhang, and Lan Wei (2019) "Developing Translator Competence: Understanding Trainers' Beliefs and Training Practices", *The Interpreter and Translator Trainer* 13(3): 233–54, doi: 10.1080/1750399X.2019.1656406.

Yan, Xiu Jackie and Wang Honghua (2012) "Second Language Writing Anxiety and Translation: Performance in a Hong Kong Tertiary Translation Class", *The Interpreter and Translator Trainer* 6(2): 171–94, doi: 10.1080/13556509.2012.10798835.

Yoon, Hyunsook and Alan Hirvela (2004) "ESL Student Attitudes toward Corpus Use in L2 Writing", *Journal of Second Language Writing* 13(4): 257–83, doi: 10.1016/j.jslw.2004.06.002

Zanettin, Federico (2002) "Corpora in Translation Practice", paper presented at the Workshop of Language Resources for Translation Work and Research, available at www.lrec-conf.org/proceedings/lrec2002/pdf/ws8.pdf

7 What Is an "Ideal" Home?
A Multimodal Discourse Analysis of the Housing Names and TV Advertisements in Hong Kong

Lam Yee Man, Lam Shu Yan, and Ng Kwan-kwan

Introduction

In modern society, a home can be a commercial product; conflating home and house, a home is "sold" and can be acquired. In Hong Kong, house acquisition is one of the status markers in the society. As house acquisition is extremely and infamously difficult, a house owner has long been defined socially as "the successful", the "elites" in Hong Kong.[1] In this chapter, we are concerned with how developers employ language and visual discourses to construct an ideal home for the "elites" in Hong Kong. As all housing in Hong Kong is named, we are interested in investigating how "home" is semioticised to demonstrate this elite identity and the ideology behind it. With the help of the computer program Python and taking a multimodal discourse analysis approach, this chapter examines signs, namely the unique names and promotional videos of the housing estates in Hong Kong from the 1980s and 2020s,[2] and studies how an elite identity is constructed and ideologies unwittingly manifested through the acquisition of a "home". We have found that "imperial flavour" is the increasing trend in private housing names—to be an elite is to be noble and king-like. Seeing that wealth disparity is worsening in Hong Kong, housing names, in fact, are on the one hand celebrating the success of the home owners and on the other hand unceasingly glorifying the elite identity, promoting alienation and a decades-long naturalization of the social hierarchy in Hong Kong.

Housing and Its Name in Hong Kong

Characteristically, all Hong Kong housing is named, be it public rental housing,[3] subsidized housing,[4] or private housing. With very few exceptions, almost all housing in Hong Kong takes the form of high-rise apartment buildings, and, if the apartment buildings are developed and belong together as an estate, they are given a common name. Hence, be it an individual apartment building or a group of apartment buildings forming an estate, they are all named by the

DOI: 10.4324/9781003376491-8

developers. It is these names, instead of the street names, which are used for daily communication, and these are the names, together with their respective TV advertisements, that are studied in this chapter.

What also characterizes Hong Kong housing is its high price. Hong Kong has long been renowned for being hugely densely populated: 7.5 million people are living in only 78 square kilometres in Hong Kong (Planning Department, 2018).[5] Although public housing for rent and subsidized flats for sale are available in Hong Kong, demand for private housing never stops increasing. There are three reasons. First, families that earn an income over the upper-income limit of public housing can only look for flats in the private property market. Second, the rapid increase in the number of domestic households (from 2.23 million in 2006 to 2.51 million in 2016), together with a decline in household size (from 3.0 in 2006 to 2.8 in 2016), leads to an enormous demand for private housing in Hong Kong (Census and Statistics Department, 2016). Third, the land supply is limited due to Hong Kong's hilly landscape. These all result in the continuous supply falling short of demand in private housing in Hong Kong,[6] making private housing unaffordably high-priced.[7] Subsequently, the housing itself becomes a sumptuous product in Hong Kong, and those acquiring private housing are defined socially as "smart" and "successful" (Cheng, 2001), the elites (Forrest and Lee, 2004), or, according to developers, the "noble class" (more on this later in the chapter).

Elite here refers more to a *feeling*, to an identity, than to the objective measurement of wealth (Thurlow and Jaworski, 2010). Just like Bourdieu's taste, "elite" is a social status, an identity which is processual, which has to be enacted and maintained through the acquisition and display of the "right" taste, sign, and the performance of the "correct" action. Language and discourse, according to Thurlow and Jaworski (2010), are two of the means through which the status of the elite is constructed and performed. As shown in this case study, the elite identity is embedded in private home owning. What this chapter aims to do is to investigate how a home is semioticised, how an elite identity is constructed through different rhetorical strategies and semiotic framing in relation to the notion of home (Simmel 1997; Entrikin, 1991; Johnstone, 2004; Netto, 2011; Liu, 2010), how the names are "invested by ideology" (Fairclough, 1995, p. 73), and, most importantly, what these ideologies are.

Home is a physical space to which one is emotionally attached (Porteous, 1976; Dovey, 1985), a place where one feels safe and secure (Dupuis and Thorns, 1996), a place where one feels autonomous (Saunders, 1990), a place which will affect our physical and mental health (Fonagy, 1999; George and West, 1999). Home is one of the major domestic units where social relations are constituted (Saunders and Williams, 1988). Home is more than a physical space; it is a space inscribed with personal and social meanings and relations (Rapoport, 1995; Easthope, 2004). Home constructs or affects a part of who we are (Marcus, 1995; Netto, 2011; Liu, 2010). And a "good" home is a mixture of all these elements: a place of connections and comfort (Fozdar and Hartley, 2014). However, beneath the construction of a home is also the notion of power and the politics of identity: not only does the way owners maintain their homes demonstrate how they define and reveal their positions in society, but the way home is normatively defined may also marginalize certain groups of people in society (Blunt and Dowling, 2006). For instance, the home of the eccentric and the homosexual (e.g. Fortier, 2003), the home of migrants (e.g. de Carvalho, 2003), and the home of refugees (e.g. Kale, Kindon, and Stupples, 2018) may question the normativity of what we usually call "home". Even the acquisition of a house as a home can be political: Although the identity of a "house owner" demonstrates one's success in the game of capitalism, the house owner submitting themselves to mortgage repayments and hence limiting revolutionary possibilities is something the authority is willing to see (Harvey, 1978). Home, in short, is a contested concept; it is both social and ideological.

Ideology refers to the system of ideas that legitimates the prevalence and reproduction of the existing power structure (Fairclough, 2001). As the system of ideas often appears in the form of "common sense", or a consensus that people are unaware of, the power which the dominant group holds is reproduced and exercised in the society in a natural and unnoticed way (Fairclough, 2001). Ideology is hence tied to power and domination, with language being "a material form of ideology" (Fairclough, 1995, p. 73). What this article aims to do is to "denaturalize" the implicit ideology (Fairclough, 1995) and to unveil the dominant ideology implied in the housing names and advertisements in Hong Kong.

Because of the limited scope of this chapter, we will focus primarily on the names and representations of housing. Names and TV advertisements are the representations *of* and *about* housing; these are the representations that show how the developers define their housing. Thus studying these names and representations will suffice to give us a comprehensive overview of the construction and manifestation of the ideology and the elite identity through home owning in this case study.

Methodology

This chapter has conducted a multimodal discourse analysis; we studied all housing names in Hong Kong from the 1980s to 2020s and their respective TV advertisements. All the names were obtained from government documents "Names of Buildings 2019" (Volumes 1[8] and 2[9]). This all-inclusive list, with the help of the computer program Python, was then cleaned and refined by (1) deleting the nonresidential buildings in Hong Kong[10]; (2) comparing the list with the "major housing estates list" issued by the government;[11] (3) comparing the list with the public and private housing lists in Wikipedia.[12] After the cleaning, there were a total of 1,101 items, including public housing, subsidized housing, and private housing.[13] The data was then counted and ranked according to the number of occurrences (Table 7.1). Since Python can

Table 7.1 Proper Name (Single Word)

	1980s	1990s	2000s	2010s	2020s
Public	田 (3), 葵 (3), 興 (2), 愛 (2), 東 (2), 順 (2), 盛 (2)	翠 (4), 華 (4), 長 (4), 景 (4)	田 (6), 東 (5), 天 (5), 安 (4), 慈 (4)	彩 (4), 逸 (3), 天 (3)	灣 (7), 朗 (5), 華 (4), 明 (4)
Subsidized	漁 (1), 暉 (1), 順, 緻, 穗 禾, 愉, 城	翠 (9), 康 (7), 麗 (7)	富 (8), 明 (8), 東 (6), 康 (6)	東 (2), 濤 (2), 盈 (2), 嘉 (2), 峰 (2), 油 (2), 翠 (2), 葵 (2)	德 (5), 山 (4), 東 (4), 啟 (3), 翠 (3), 綠 (3), 錦 (3)
Private	康 (5), 景 (5), 威 (5), 華 (4), 碧 (4), 富 (4)	華 (9), 翠 (8), 麗 (7), 樂 (6), 都 (6), 新 (6), 怡 (6), 荃 (6)	豪 (18), 海 (16), 景 (16), 灣 (16), 翠 (12), 帝 (12), 峰 (10), 軒 (7), 龍 (6), 御 (4)	山 (15), 庭 (14), 豪 (12), 御 (11), 海 (10), 匯 (10), 逸 (7), 君 (7), 凱 (6), 皇 (5), 龍 (5), 帝 (4)	海 (18), 御 (16), 尚 (15), 匯 (15), 薈 (11)

Note: The number of occurrences of the word is stated in the bracket. Words related to nature are set in black bold, words to supremacy, grey bold.

only calculate the names as separate individual words, the housing names were also read manually as compounds and were grouped into seven different categories (Table 7.2). Whereas the use of Python enables us to get an overview of the trend and development of the housing names through the decades, the housing TV advertisements provide crucial qualitative proof. Text analysis was then performed to investigate how an elite identity is constructed through the acquisition of a home.

In what follows, we will present the results of the quantitative analysis, discuss and explain the trends of the housing names, examine the TV advertisements, and conclude the chapter by unveiling the implied ideology. Noteworthily, most housing names are composed of two parts: a proper name and a generic noun. And as the names studied in this article are mainly Chinese characters, the Cantonese pinyin is written in quotation marks with its literal translation in the bracket. For instance, the public housing "Waan Ceoi Tsuen" (環翠邨) is composed of the proper name "Waan Ceoi" (環翠, surrounded by greenery) and the generic noun "Tsuen" (邨, estate).[14] As housing English names may not be the direct translation of their Chinese counterparts—the English names may carry different meanings (to be discussed)—it is beyond the scope of this

What Is an "Ideal" Home? 97

Table 7.2 Reading the Proper Name as a Compound

	1980s	1990s	2000s	2010s	2020s
Geographical Location					
Public	23 (58%)	23 (35%)	15 (33%)	9 (30%)	0 (0%)[44]
Subsidized	0 (0%)	4 (5%)	11 (12%)	8 (33%)	42 (64%)
Private	14 (21%)	28 (18%)	17 (11%)	13 (7%)	0 (0%)
Prosperity of the Nation					
Public	4 (10%)	9 (14%)	2 (4%)	0 (0%)	0 (0%)
Subsidized	0 (0%)	3 (4%)	3 (3%)	0 (0%)	0 (0%)
Private	5 (7%)	8 (5%)	1 (1%)	1 (1%)	0 (0%)
Wish					
Public	8 (20%)	16 (24%)	20 (43%)	11 (37%)	11 (61%)
Subsidized	2 (50%)	35 (47%)	48 (51%)	4 (17%)	6 (9%)
Private	26 (39%)	40 (26%)	25 (16%)	14 (7%)	0 (0%)
Nature					
Public	17 (43%)	27 (41%)	15 (33%)	11 (37%)	6 (33%)
Subsidized	2 (50%)	32 (43%)	25 (26%)	16 (67%)	28 (42%)
Private	19 (28%)	80 (52%)	86 (54%)	97 (51%)	0 (0%)
Imperial Supremacy					
Public	0 (0%)	0 (0%)	1 (2%)	5 (17%)	1 (6%)
Subsidized	0 (0%)	3 (4%)	12 (13%)	3 (13%)	2 (3%)
Private	2 (3%)	9 (6%)	30 (19%)	58 (31%)	69 (100%)
Others*					
Public	0 (0%)	1 (2%)	0 (0%)	0 (0%)	0 (0%)
Subsidized	0 (0%)	0 (0%)	0 (0%)	0 (0%)	1 (2%)
Private	6 (9%)	12 (8%)	14 (9%)	16 (8%)	0 (0%)

Note: According to their respective meanings, manually we have classified them into six categories. We have calculated the trend percentage (listed in the bracket) by using the total number of that type of housing built in that decade as the dividend. Note that as some names which indicate their geographical location also bear a strong affiliation with nature; these names are double-counted in both "Geographical Location" and "Nature", resulting in a total percentage greater than 100.

* Names which cannot be put into these five categories all fall into the "others" category, including buildings named after a person (e.g. "Jat Min Chuen"(乙明邨) and buildings denoting a sense of "newness" (e.g. "San Dou Gwong Ceong" (新都廣場/new city plaza).

chapter to examine all English names in this study. This chapter will focus mostly on the Chinese names and the TV representations of the housings.

Trends and Differences of the Housing Names

As shown in Table 7.2, there is a clear trend in public housing names of characterizing the geographical location (>30% from the 1980s to 2010s) and the natural elements (>30% thorough the decades). For the geographical location, public housing may form the proper name by adapting one word from its geographical location. For instance, "Gwong Tin Tsuen" (廣田邨,

extensive field) is a blending of the location-related word "Tin" (田, field) from "Lam Tin" (location) with "Gwong" (廣, extensive); likewise, "Coi Wan Tsuen" (彩雲邨, colourful cloud) blends the location-related word "Coi" (彩, colourful) from "Coi Hung" (location) with "Wan" (雲, cloud). Hence these names are denotative—by looking at the location-related words, the locals can guess easily the location of the housing. Further, it is also very common for public housing to form proper name by including the natural elements; apart from the "field" and "cloud" mentioned, other examples include: "Tin" (田, field), "Kwai" (葵, sunflower), "Ceoi" (翠, evergreen), "Tung" (東, east), "Coi" (彩, colour), "Wan" (灣, bay), as shown in Table 7.1.

Slightly different from public housing, although natural elements also take a dominant role in subsidized housing (except in the 2000s, ⩾50% through the decades, see Table 7.2), such as "Jyu" (漁, fishing), "Fai" (暉, sunshine), "Tou" (濤, wave), and "Shan" (山, hill). As shown in Table 7.1, the wish for one's well-being also tops the list (about 50% from the 1980s to the 2000s): "Hong" (康, health), "Fu" (富, wealth), and "Dak" (德, virtue). The proper name of subsidized housing has a similar but slightly different focus, when compared with the public ones.

Comparatively, although naming the private housing with natural elements and the wish for one's well-being is also common, especially in the 1980s and 1990s (about 50% over the decades, see Table 7.2), e.g. "Hong" (康, health) and "Ceoi" (翠, evergreen), a new and distinctive trend emerges from the 2000s onwards: imperial supremacy (from 19% in the 2000s increased to 100% in the 2020s). Examples of imperial supremacy include "Hou" (豪, commander in chief); "Ting" (庭, the imperial court); "Jyu" (御, rule or royal), and "Soeng" (尚, rule and respect). Therefore, by reading the name "Dai Ging Waan" (帝景灣, King's viewbay), which is frequently used with the generic noun "Waan" for private housing with its imperial flavour, one could be quite certain that it is private housing.

Summarizing Tables 7.1 and 7.2, whereas public housing is named mainly in relation to its geographical location with an added affiliation with nature, subsidized housing names express wishes and virtues; through the decades, private housing has gradually emphasized a sense of superiority, a sense of eliteness. Although nature is a vital element in all three types of housing, the images it inspires and the connotations it invokes are different, as we are going to see.

Exclusivity: Neighbouring Nature

Nature is one of the most frequent elements in housing names in Hong Kong; almost every one out of three contains a natural element. Noteworthily, the words used in subsidized housing and their connotations are different from those used in the private ones. The words used in subsidized housing, for instance, "Ceoi" (翠, green), "Tou" (濤, wave), "San" (山, hill), "Dung" (東, east) are different from those used in private ones, "Hoi" (海, sea), "Ging" (景, view), "Waan" (灣, bay), "Wui" (薈, luxuriant vegetation) in that the latter seems to suggest a stronger affiliation with the sea. This can be further exemplified by a study on the compound level. "San Ceoi Tsuen" (新翠邨, new green), "Choi Fung Yuen" (彩峰苑, colourful summit), "Paak Wai Yuen" (柏蕙苑, cypress and orchid), and "Mei Cung Yun" (美松苑, beautiful pine tree). All these subsidized housing names suggest a relation with nature on a rather micro level, connoting more of the beauty of a tree or a plant. On the contrary, the compounds in private housing, especially those after the 2000s, as weird as it may seem, not only connote a close relationship with the sea but also excite a relatively more panoramic and scenic vision: "Zeon Jyun" (峻弦, high crescent), "Bik Hoi Laam Tin" (碧海藍天, blue sea and clear sky), "Hoi Din Geoi" (海典居, sea scripture), "Hoeng Dou" (香島, fragrant island), "Hoi Tou Geoi" (海濤居, sea wave), "Hoi Ji Bun Dou" (海怡半島, pleasant peninsular), "Jyu Hoi Waan" (月海灣, sea under the moon), and "Laam Cing Waan" (藍澄灣, blue clear bay).

The popularity of natural elements in housing names should not be too surprising. In traditional Chinese culture, neighbouring nature has long been regarded as an essential of an ideal home.[15] Research has also demonstrated that encountering nature is mentally, psychologically, and physically beneficial to humans.[16] It is, therefore, not surprising that, for many, constituting an "ideal" home is to make a neighbour of nature. However, the housing names are in fact aspirational. For instance, the two most frequently used generic nouns "Yun" (苑, garden) and "Faa Yun" (花園, garden), used in subsidized and private housing, both signify gardens in Chinese. Similarly in private housing, the most frequently used generic nouns in private housing after the 2000s— "Waan" (灣, bay), "Fung" (峰, summit) both bear an affiliation with nature. However, this affiliation with nature should not be overstated, for these nature-related lexicons very often do not denote their usual signifiers in reality. For instance, although "Tsuen Wan Faa Yuen" (荃灣花園, Tusen Wan Garden) may literally denote a beautiful garden, the private housing is in fact surrounded primarily by concrete buildings with limited greenery (Figures 7.1 and 7.2). Likewise, the word "Garden" in "Lung Fai Faa Yuen" (龍暉花園, Radiant Dragon Garden) only denotes "garden" in a wishful, aspirational way, for this private housing is surrounded by other concrete high-rise buildings in the area (see Figures 7.3 and 7.4); a real, green garden is absent. Likewise, the use of "Waan" (灣, bay) and "Fung" (峰, summit) may not denote that the housing is located by a bay or at a summit; the wording may just be symbolic and aspirational. Another example, the name "Bik Hoi Laam Tin" (碧海藍天, blue sea and clear sky) may not denote housing by the blue sea and a clear sky. Seeing that Hong Kong's air quality falls short of the international standard (Brajer, Mead and Xiao, 2006), whether a clear blue sky and sea can be seen is highly doubtful. Perhaps, the housing names (and TV advertisements) being

Figure 7.1 Tsuen Wan Garden, Google map street view (© 2023).

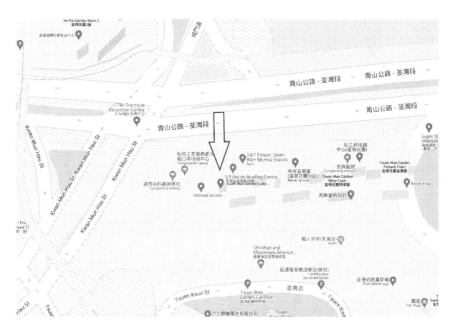

Figure 7.2 Tsuen Wan Garden, Google map (© 2023).

What Is an "Ideal" Home? 101

Figure 7.3 Dragonfair Garden, Google map street view (© 2023).

aspirational can be best evidenced by a statement made in all private housing promotional materials published after 2013 April:

> The photographs, images, drawings or sketches shown in this advertisement/promotional material represent an artist's impression of the development concerned only. They are not drawn to scale and/or may have been edited and processed with computerized imaging techniques. Prospective purchasers should make reference to the sales brochure for details of the development.

The wish for neighbouring nature can be easily understood. Due to the rapid urbanization and inadequate green space in Hong Kong (Tian and Tao, 2012), nature is disappearing; neighbouring nature becomes very difficult if not impossible to find. As such, the natural elements in the name may provide mental comfort or some remedy for the disappearing and limited nature in the urban area. However, it could be also because, given nature is so precious in Hong Kong, nature becomes a product, an asset adding value to private housings (Cheng, 2001). This can explain why the private housing names are connoting a closer connection with the sea— a spacious apartment with a sea view is one of the "standards" of luxurious private housing.[17] Nature is used as an important product to construct the prestigious new middle-class identity for residents/potential buyers. And the key to this is exclusivity.

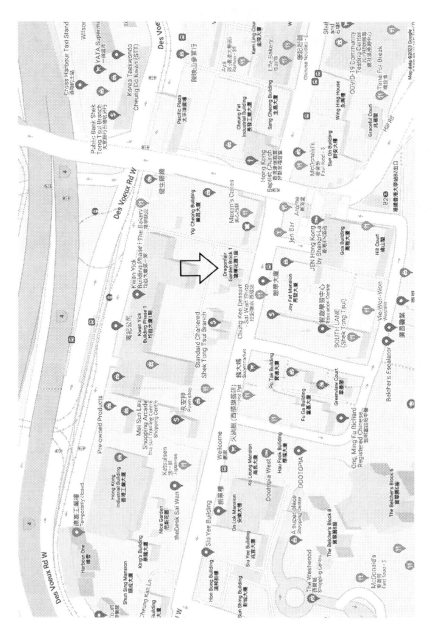

Figure 7.4 Dragonfair Graden, Google map (© 2023).

What Is an "Ideal" Home? 103

As shown in the TV advertisement of "Hoi Din Geoi" (海典居, Sea Scripture)[18] in the 1990s,[19] the advertisement was rather simple and straightforward: a young couple is enjoying a panoramic sea view from their balcony. In this advertisement, first, the sea is clear and uncontaminated (which is rare in Hong Kong); second, apparently, nobody is around but only the young couple, connoting the privatization of the sea. Here, nature (sea) is introduced as rare property, as a privilege enjoyed by the very few. This exclusivity is further underscored by the use of possessive pronouns and the creation of a sense of intimacy. The video ends with this statement, "with the sky being your companion, you dance with the sea—what an ideal world".[20] Emphasized here is the exclusivity brought out not only by the possessive pronoun "your" but also by an activity: Nature is now an anthropomorphized dancing partner implying that you and nature are a pair, being together, and connoting a sense of intimacy. Similar examples are many; in other TV advertisements, nature is represented as an "honorable neighbour",[21] *your* lover,[22] similarly connoting a private intimacy with nature. Note that similar stereotypical elements and semiotic resources can still be found in the 2010s. For instance, in the TV advertisement of "Hoeng Dou" (香島, Fragrant Island), a young family enjoying activities in the uncontaminated sea and unpopulated field is again connoting the privatization of nature. The concluding statement is especially telling: "I am so glad to be *so close* with you, nature. This is *my* nature".[23] Again, underscored here are the two notions: exclusivity and intimacy. These two notions turn nature from a common good to a rivalrous good, implying that nature is now a product that may be consumed only by a single user. It is through this exclusivity and intimacy that the prestigious, elite identity of the residents/home owners is constructed and manifested.

Exclusivity: The Chinese Imperials

Exclusivity is also employed in private housings names with imperial flavour. As shown in Table 7.2, imperial supremacy is one of the major themes in private housing. Although Chinese imperial-rule-related words are occasionally used in public and subsidized housing (see Table 7.2), the words chosen are different from those used in private housing. In public and subsidized housing, quite often, it is the word "Lung" (龍, dragon), symbolizing the emperor in ancient China, which was used, such as "Lung Jat Tsuen" (龍逸邨, dragon at ease) and "Lung Hin Yun" (龍軒苑, room for dragon). In comparison, the words employed in private housing names are diverse; besides the word "Lung", other imperial-related words are also used: "Dai"(帝, king) in Dai Fung, "Wong Din" (帝峯．皇殿, the king summit, the royal palace);[24] "Ting" (庭, royal court) in "Ji Ting Geoi" (怡庭居, pleasant royal garden);[25] "Jyu" (御, royal) in "Jyu Wong Toi" (御皇臺, royal, imperial), "Gwan"(君, king) in "Gwan Jyut Hin" (君悅軒, pleasant king); "Hoi" (凱, triumph) in "Hoi Jyu" (凱譽, triumph). However, what marks the difference between private and public housing is the use of rare and very uncommon words in the former; the words used in private housings are so rarely used that they are tested as difficult questions in an online Cantonese knowledge competition.[26] Most of these are words that are with the Chinese Radical (i.e. the indexing component in the Chinese character) "玉" (jade).[27] As jade was one of the favourites of the emperors in ancient China, words with the radical jade (玉) are related to the imperial order. Perhaps this can be best exemplified by the name "Gau Lung Saan" (玖瓏山); instead of using the widely used compound "九龍", "玖瓏" (same pronunciation but with the radical jade) is used as the housing name.[28] In addition, the employment of the uncommon imperial-related words in private housing names is intended to create a sense of superiority. The uncommonly used words in the private names successfully create a sense of distance and difference—that they are uncommon, unique, rare, and hence exclusive.

Exclusivity: The Western Imperials

The imperial-flavoured private housing names are lexically related to the imperials in ancient China. However, from the TV representations published by the developers, the "imperials" are, connotatively, very "Western". This may be because the phrase "middle class" no longer suffices to articulate the superior social status of this group of people, they have to be represented by different and stronger words. Since the 2000s, the developers start to affiliate private housing with the word "noble" in the advertisements,[29] and it is "noble" in the European sense. Examples are plentiful in the TV advertisements: young ladies and gentlemen, some even in their nineteenth-century European costumes or modern nightgowns, attending parties[30] and concerts,[31] riding horses in the wild field,[32] riding in a British-like royal horse cart,[33] playing golf,[34] reading in a European-style library,[35] strolling in a European imperial court,[36] shopping on a Milano-like street,[37] etc. These are all uncommon activities in Hong Kong but typical imaginations of the European style of prestigious living. These are all status markers (Cheung and Ma, 2005). This noble class identity can best be illustrated in the TV advertisement of "Dai Fung, Wong Din" (帝峯．皇殿, the king summit, the royal palace). The advertisement introduces how the housing is related to the essence of the House of Romanov (Russia), including some of the clubhouse's design and architecture and the treasure of Queen of St Petersburg's Legacies acquired from Sotheby's Auction; the housing is described as the "palace in the city".[38] Similar examples are abundant: In the TV advertisement, the private housing "Wong Fu Shan" (皇府山, Noble Hill) features the tradition, lifestyle, and leisure of the European noble class;[39] Ging Si Baak Shan (京士柏山, King's Park Hill), using Baroque music as the background music, highlights the European-style garden and gateway, having a demeanour of a king;[40] Hoi Syun Mun's (凱旋門, Arc De Triomphe) advertisement introduces the inspiration of the design of the housing stemming from the Arc de Triomphe in Paris.[41]

All these borrowings from European tradition and culture are intended to deliver a sense of exclusivity and a sense of superiority; such private housing is exclusively for kings and nobles. They create fantasies of a different identity (Featherstone, 1990; Arnould and Price, 1993), the "I am = what I have and what I consume" (Fromm, 1975: 36); more precisely, "I am = where I live" (Liu, 2010). The housing is intended to show the "bestowed privileged status, tremendous achievement"[42] of the residents, the new noble, the elite in Hong Kong. This constant and increased need to demonstrate one's eliteness through the commercial product of a "home", especially after the 2000s, can be explained. Despite the fall during the Asian Financial Crisis in 1997 and 1998, the housing price continued to rise for decades, reaching its new peak in the 2010s (Li, 2017). Just as in other Asian countries, because of low taxes and the small government, a new middle class emerged after the Asian Financial Crisis. This new middle class needed means to demonstrate their identity and social status—and did so through their consumption habits (Chua, 2000) and house acquisition in Hong Kong (Cheng, 2001). And the particular reliance on the European context and lifestyle is intended to create a fantasy of a lifestyle and identity markedly *different* from those living in other types of housing; *differences* which can demonstrate the identity of this new noble class—they are the ones who know how to live differently from the "ordinary" life in Hong Kong. As Hong Kong society has become more affluent (Chan, 2000), the original class marker, such as the choice of food and fashion (Bourdieu 2010) or being a private housing owner (Marcuse, 1975) may no longer be sufficient to demonstrate one's real class status (Cheng, 2001; Jenkins et al., 2015). Developers therefore are trying to create a new class marker: The housing names themselves should disclose the social status of the buyers/residents, names which distinguish them from public and subsidized housing, names which emphasize more "highness" and differences, power, and imperial supremacy. Therefore, what is sold is not only the use and exchange value of a private home but also its symbolic value (Baudrillard, 1998)— here, eliteness and supremacy.

Distancing the Poor

Behind the notion of exclusivity and the sense of eliteness is an intention to alienate oneself from the public. An ideal home for the elite alienates them from the public. This can be further elucidated in the example of 8 Waterloo. This private housing, named 8 Waterloo,[43] is located at the junction of Waterloo Road, Portland Street, and Shanghai Street. Seeing that Portland Street and Shanghai Street are infamous for the red light district and a district of a mixed bunch of good and bad (that is, the "lower class"), the developer is apparently trying to shift the emphasis away from the infamous Portland Street and Shanghai Street and toward Waterloo Road. One of the major thoroughfares in Hong Kong, Waterloo Road connects Yaumatei (where the private estate is located) with Kowloon Tong (where many reputable and luxurious houses, apartments, and famous schools are located). Seemingly, the emphasis on Waterloo Road in the housing name is an effort to create a vision of "neighbouring" with privileged groups—even though 8 Waterloo is in reality about 3 kilometres away from Kowloon Tong, where the privileged groups are located—while maintaining a virtual distance from the "bad" and the poor of Shanghai Street and Portland Street.

The sense of alienation and difference can be further evinced by the English names of the private housings. English and European languages are used to give the product a high feeling of status (Jaworski and Yeung, 2010). Whereas some of the housings are translated literally, especially public housing (for instance, "Lai Tak Tsuen" (勵德邨) as Lak Tak Estate), private housing names are translated differently. Quite often, they employ towers, buildings, cities in other countries in naming the private housings: in "English", the private housing "Sing Tai" (星堤, Star) is named "Avignon", a commune in France. Likewise, "King Tin Bun Dou" (擎天半島, sky-upholding) is translated as "Sorrento" (a town in Italy), "Bik Tai Bun Dou" (碧堤半島, the dike of jade) as "Bellagio" (a resort in Las Vegas), "Laam Ngon" (嵐岸, Mountain mist) as "Sausalito" (a city in California), and "Dai King Waan" (帝景灣, king's view) as "Corinthia by the sea" (a city in Greece). Apparently, the further away from the "local"/ "general public" , be it only a virtual distance, the more prestigious they are. All these show that private housing names are used to create difference, distance, and alienation. They are intended to demonstrate one's social status, to exclude oneself from the mass public, and to be included in the list of the privileged. A home for the elite ought to be different.

The social importance of this difference and alienation and the significance of this manifestation of elite identity may not be fully articulated unless the contradiction in the society is also spelled out. Hong Kong is equally infamous for its wealth gap, with its Gini coefficient index worse than other economies including Singapore, the United States, the United Kingdom, Australia, and Canada (Oxfam, 2018). It is under this context that the glorification and the reproduction of the elite identity are problematic. With the wealth disparity worsening (Lam and Kwan, 2020), one is encouraged to pursue and manifest publicly their superior social status. With over 1.3 million people living in poverty (Oxfam, 2018), some of whom are even "sleeping with suitcases, cooking next to the toilet" (Lok, 2016), the "elite" is encouraged to imagine life in a Western/European imperial context. If advertising is a mirror reflecting the cultural values of a place (Polley 1987), underneath this notion of an "ideal" home is, from the 1980s to the 2020s, a *continuous* indulgence of superiority and domination, an encouragement of alienation, an uninterrupted glorification, reproduction, and a naturalization of the social hierarchy in Hong Kong. Home, seen by many as a place of comfort, becomes another commercial product to manifest their superior identity and to promote and naturalize alienation and domination.

Notes

1 Individuals are encouraged to acquire a house as home in Hong Kong, see Cheng 2001.
2 Only private housing TV advertisements in Hong Kong are studied; public and home ownership housing were not promoted through TV.
3 I.e. public rental housing provided by the Hong Kong government.
4 I.e. public housing sold by the Hong Kong government at a relatively affordable price.
5 Among the residential housing in Hong Kong, 29.1% of people are living in public rented housing, 15.5% in subsidized sale flats, and 54.8% in private housing (Transport and Housing Bureau, 2019).
6 This can be exemplified by the housing price index, which in Hong Kong has sharply increased from 138 in 2010 to 379 in 2020 (Rating and Valuation Department, 2020).
7 Shown in Demographia International Housing Affordability Survey (Cox and Pavletich, 2019), Hong Kong's housing tops the list as the most "severely unaffordable" with a median multiple of 20.9 in 2018, compared to 12.6 of Vancouver and 11.7 of Sydney.
8 "Names of buildings" (2019), Volume 1, www.rvd.gov.hk/doc/en/urban.pdf
9 "Names of buildings" (2019), Volume 2, www.rvd.gov.hk/doc/en/nt.pdf
10 In Chinese, a building's name is composed of a proper name and a generic noun, with the latter indicating the type of the building. For instance, "香港仔工業大廈 (Aberdeen Industrial Building)" is composed of the proper noun "Aberdeen" ("香港仔") and the generic words "Industrial building (工業大廈)". It is according to this principle that we deleted items with the following generic words in the list, including schools (學校, 中學, 幼稚園), hotels (酒店), commercial and industrial buildings (工業大廈, 廠, 商業大廈, 商業中心), buildings providing government/public services (署, 垃圾站, 市政大廈, 體育館, 醫院, 市場, 街市), religious buildings (僧舍, 教堂, 會堂), nongovernmental organizations (協會, 服務中心), elderly homes (安老院), and private clubhouses and shopping malls (中心, 會所, 廣場).
11 "Major housing estates", Population by census 2016, www.bycensus2016.gov.hk/en/bc-dp-major-hosing-estates.html
12 "List of public housing estates in Hong Kong", https://en.wikipedia.org/wiki/List_of_public_housing_estates_in_Hong_Kong; "Private housing estates in Hong Kong", https://en.wikipedia.org/wiki/Private_housing_estates_in_Hong_Kong
13 Although buildings to be built in the 2020s are included in this study, the list is not exhaustive and is subject to change, be it the numbers of buildings to be built or the names of these buildings.
14 The Chinese Pinyin is transcribed with reference to "S. L. Wong's *A Chinese Syllabary Pronounced according to the Dialect of Canton*. See http://humanum.arts.cuhk.edu.hk/Lexis/Canton/
15 For instance, as shown in "The Crude House" (陋室銘) by Liu Yu Xi (劉禹錫) from the Tang dynasty, the house is described as neighboring both mountain and water (lake) and as being surrounded by greeneries. In "Phoebe Zhennan Being Uprooted by Wind and Rain" (枏樹為風雨所拔歎), Du Fu (杜甫), from the Tang dynasty, explains in the poem that the old tree and sea are the criteria for his ideal home.

16 Natural elements are said to encourage social interactions of a community (Coley, Sullivan, and Kuo, 1997); nature improves children's attention capacity (Wells, 2000); a view of/exposure to nature facilitates our positive emotions (Bratman, Hamilton, and Daily, 2012; Ryan et al., 2010; Aspinall et al., 2015); lowers our stress level (Atchley, Strayer, and Atchley, 2012); improves one's self-reported well-being, be it at home (White et al., 2013), at leisure (Lehto, 2013), or at work (Gilchrist, Brown, and Montarzino, 2015).

17 Midland Realty, 2020. "【全港】想跟大海做鄰居 香港多少錢有？" (How much do you need if you want to live by the sea?) May 29, www.midland.com.hk/property-news/%E6%9C%80%E6%96%B0/%E3%80%90%E5%85%A8%E6%B8%AF%E3%80%91%E6%83%B3%E8%B7%9F%E5%A4%A7%E6%B5%B7%E5%81%9A%E9%84%B0%E5%B1%85-%E9%A6%99%E6%B8%AF%E5%A4%9A%E5%B0%91%E9%8C%A2%E6%9C%89%EF%BC%9F/. Accessed June 10 2021.

18 香港經典廣告(1998)，海典居 Villa Oceania. www.youtube.com/watch?v=lNC2HRKwrBQ Accessed August 6, 2020.

19 One more example from "Hoi Tou Geoi" (海濤居, sea wave) is another example. Also featuring the lives of a young family, enjoying the massive sea view, the video ends with the statement: "[T]he sky and the sea merges, how free and tranquilizing; magnificent view, great relaxation, let's merge with nature —your prestigious housing" (「海天一色，自由閒適；無盡景色，無盡寫意；與自然融而為一；高級住宅」, translated by the author). From: 電視廣告 1892 馬鞍山 海濤居, www.youtube.com/watch?v=zCGOR-ZPzH8 Accessed October 2020.

20 Original: 「世外桃源，與天為伴，與海共舞」(translated by the author).

21 香港廣告 (2017) 星漣海，SEANORAMA, www.youtube.com/watch?v=uHv7FRIFWMM. Accessed August 6, 2020.

22 香港廣告, (2017) 意花園，CRESCENDO, www.youtube.com/watch?v=zOrBTQCIQE8. Accessed August 6, 2020.

23 香港廣告 (2016), ISLAND GARDEN, 香島 www.youtube.com/watch?v=omsYyXBR9c4. Accessed August 6, 2020. Original quotation,「每朝早可以同你咁親密，多謝你，人自然。(This is my nature」, translated by the author; emphasis added).

24 Another example is "Dai Wui Hou Ting" (帝匯豪庭, the place where kings gather).

25 Another example is "Dai Ting Jyun" (帝庭園, Royal garden)

26 See WHIZOO擂台 – HK香港粵語大字典 (literal translation: WHIZOO competition—HK Cantonese dictionary), www.youtube.com/watch?feature=share&v=d78ehka7wxM&app=desktop

27 For example, 尚璟、瑧環、瑧蓺、璥珀、瑧璥、珵華、天璽、玖瓏山.

28 Apart from these words with the radical jade, other words related to the imperial order are also used, such as "Hou Tin" (皓畋) and "Hon Lam Fung" (翰林峰). The former refers to royal hunting; the latter denotes the place where the government officials stayed.

29 "Noble" is a word very frequently employed in the private housing TV advertisements since 2000s.

30 TV advertisement香港經典廣告 (2006), 城中駅 Le Point-2, www.youtube.com/watch?v=00nPernsFsE. Accessed August 8, 2020.

31 TV advertisement 富甲半山 1' 廣告, www.youtube.com/watch?v=SUbQcgefCtQ&list=PL7359037B0FE3FC56&index=4. Accessed August 8, 2020.

32 TV advertisement香港經典廣告 (2008)御龍山, The Palazzo, www.youtube.com/watch?v=Qx7GfSrWTXg. Accessed August 8, 2020.
33 TV advertisement香港經典廣告 (2006), 比華利山別墅, The Beverly Hills-2, www.youtube.com/watch?v=FESi69g0ZGc. Accessed August 8, 2020.
34 TV advertisement 香港經典廣告 (2006), Royal Green 御皇庭, www.youtube.com/watch?v=0xi5Nuj76JE. Accessed August 8, 2020.
35 TV advertisement 香港地產廣告：一號銀海, www.youtube.com/watch?v=CuNDMQe3DSw. Accessed August 8, 2020.
36 TV advertisement 香港經典廣告 (2000), 清水灣半島 Oscar by the Sea, www.youtube.com/watch?v=_vIytZY1R9o. Accessed August 8, 2020.
37 TV advertisement香港經典廣告 (2004), 宇晴軒, The Pacifica, www.youtube.com/watch?v=sG6lMsSUE6o. Accessed August 8, 2020.
38 TV advertisement史上最誇張最痺樓盤廣告,帝峯．皇殿, The Hermitage (2010), www.youtube.com/watch?v=-RgQK_TjT0o. Accessed August 8, 2020.
39 香港經典廣告, (2005), 皇府山, Noble Hill, www.youtube.com/watch?v=5J3Br0Wj4i0. Accessed June 4, 2021.
40 香港經典廣告, (2001), 京士柏山, King's Park Hill, www.youtube.com/watch?v=uOak4aOwU-0. Accessed June 4, 2021.
41 香港經典廣告, (2005), The Arch, 凱旋門, www.youtube.com/watch?v=Qoa8xXR5WgE. Accessed June 4. 2021.
42 TV advertisement, 富甲半山 1' 廣告, www.youtube.com/watch?v=SUbQcgefCtQ&list=PL7359037B0FE3FC56&index=4. Accessed August 8, 2020. Original quotation, 「天賦地位，超然成就」、「啟承顯赫，地位盡握」 (translated by the author).
43 This is private housing which does not come with a Chinese name.
44 As the list is compiled from the "Names of buildings" (2019), not all housings built in 2020s are included in the study here.

References

Arnould Eric, and Linda Price. (1993) "River magic: Extraordinary experience and the extended service encounter". *Journal of Consumer Research* 20(1): 24–45.

Aspinall, Peter, Panagiostis Mavros, Richard Coyne, and Jenny Roe. (2015) "The urban brain: Analysing outdoor physical activity with mobile EEG". *British Journal of Sports Medicine* 49: 272–76.

Atchley, Ruth Ann, David L. Strayer, and Paul Atchley. (2012) "Creativity in the wild: Improving creative reasoning through immersion in natural settings". *Plos One* 7(12):e51474.

Baudrillard, Jean. (1998) *The consumer society*. Thousand Oaks, CA: Sage Publications.

Blunt, Alison and Robyn Dowling. (2006) *Home*. New York, NY: Routledge.

Bourdieu, Pierre. (2010) *Distinction: A social critique of the judgment of taste*. London; New York: Routledge, Taylor & Francis Group.

Brajer, Victor, Robert W. Mead, and Feng Xiao. (2006) "Valuing the health impacts of air pollution in Hong Kong". *Journal of Asian Economics* 17(1): 85–102.

Bratman, Gregory N., J. Paul Hamilton, and Gretchen C. Daily. (2012) "The impacts of nature experience on human cognitive function and mental health". *Annals of the New York Academy of Sciences* 1249: 118–36.

Census and Statistics Department. (2016) "Domestic Households in Hong Kong" *Snapshot of the Hong Kong population*. Population by-census 2016. Available at: www.bycensus2016.gov.hk/en/Snapshot-04.html#:~:text=1.,to%202.51%20million%20in%202016. (accessed 27 August 2020).

Chan, Kam Wah. (2000) "Prosperity or inequality: Deconstructing the myth of Home Ownership in Hong Kong". *Housing Studies* 15(1): 28–43.

Cheng, Helen Hau-ling. (2001) "Consuming a dream: homes in advertisements and imagination in contemporary Hong Kong". In: Mathews, G. and Lui, T. L. (eds) *Consuming Hong Kong*. HK: Hong Kong University Press, 205–36.

Cheung, Sidney C. H., and Eric K. W. Ma. (2005). "Advertising modernity: Home, space and privacy". *Visual Anthropology*, 18(1): 65–80.

Chua, Beng-Huat. (2000) *Consumption in Asia*. London: Routledge.

Coley, Rebehkah Levine, Ming Kuo, William C. Sullivan. (1997) "Where does community grow? The social context created by nature in urban public housing". *Environment and Behavior* 29(4): 468–94.

Cox, Wendell, and Huge Pavletich. (2019) 15th Annual Demographia International Housing Affordability Survey. Available at: From www.demographia.com/dhi2019.pdf (accessed 27 August 2020).

de Carvalho, D. (2003) *Migrants and Identity in Japan and Brazil: The Nikkeijin*. Abingdon, Oxon: Taylor and Francis.

Dovey, K. (1985) "Home and homelessness". In: Altman, I. and Werner, C. M. (eds), *Home Environments*. New York and London: Plenum Press, 33–64.

Dupuis, A. and Thorns, D. (1996) "Meanings of home for older home owners". *Housing Studies* 11 (4): 485–501.

Easthope, H. (2004) A place called home. *Housing, Theory and Society* 21: 128–38.

Entrikin, J. Nicholas. (1991) *The Betweenness of Place: Towards a Geography of Modernity*. Baltimore: John Hopkins University Press.

Fairclough, Norman. (1995) *Critical Discourse Analysis: Papers in the Critical Study of Language*. London; New York: Longman.

Featherstone, Mike. (1990) "Perspectives on consumer culture". *Sociology*, 24(1): 5–22.

Fonagy, P. (1996) "Patterns of attachment, interpersonal relationships and health". In: Blane D, Brunner E and Wilkinson R (eds) *Health and Social Organisation*. London: Routledge, 125–151.

Forrest, Ray and James Lee. (2004) "Cohort effects, differential accumulation and Hong Kong's volatile housing market", *Urban Studies*, 41 (11): 2181–96.

Fortier, A. M. (2003) "Making home: Queer migrations and motions of attachment". In: Ahmed S, Castada C, Fortier AM and Sheller M (eds) *Uprootings/regroundings: Questions of Home and Migration*. Oxford: Berg Publishers, 1–20.

Fozdar, Farida and Lisa Hartley. (2014) "Housing and the creation of home for refugees in Western Australia". *Housing, Theory and Society* 31(2): 148–73.

Fromm, Erich. (1976) *To Have or to Be*. London: Abacus.

George, C. and West, M. (1999) "Developmental vs. social personality models of adult attachment and mental ill health". *British Journal of Medical Psychology* 72 (Pt3): 285–303.

Gilchrist, Kathryn, Caroline Brown, and Alicia Montarzino. (2015) "Workplace settings and wellbeing: Greenspace use and views contribute to employee wellbeing at peri-urban business sites". *Landscape and Urban Planning* 138: 32–40.

Harvey, D. (1978) "Labor, capital and class struggle around the built environment in advanced capitalist societies". In: Cox K. (ed.) *Urbanization and Conflict in Market Societies*. Chicago: Maaroufa, 265–95.
Jaworski, Adam, and Simone Yeung. (2010) "Life in the Garden of Eden: The Naming and Imagery of Residential Hong Kong" in Elana Shohamy, Eliezer Ben-Rafael, Monica Barni (Ed) *Linguistic Landscape in the City*. Bristol: Multilingual Matters, 153–81.
Johnstone, Barbara. (2004) "Place, globalization and linguistic variation". In Fought, C (ed), *Sociolinguistic Variation: Critical Reflections*. New York: Oxford University Press, pp. 65–83.
Kale, A., Kindon, S., Stupples, P. (2018) I am a New Zealand citizen now—This is my home. *Journal of Refugee Studies*, fey060.
Lam, Eric and Shawna Kwan. (2020) "Hong Kong's Yawning Wealth Gap Grows Wider Amid Pandemic". *Blomberg Quint*, July 18.
Lehto, Xinran Y. (2013) "Assessing the perceived restorative qualities of vacation destinations". *Journal of Travel Research* 52: 325–39.
Li, Sandy. (2017) "Hong Kong home prices scale new peak, 20 years after 1997 record". *South China Morning Post*, June 30.
Liu, Jing Jing. (2010) "Contact and identity: the experience of 'China Goods' in a Ghanaian Marketplace". *Journal of Community and Applied Social Psychology* 20(3): 184–201.
Lok, Elaine. (2016) "Life inside 55 sq ft: sleeping with suitcases, cooking next to the toilet and living on less than HK$300 a week". *Young Post*, 15 December.
Marcus, C. C. (1995) *House as a Mirror of Self*. Berkeley, Calif.: Conari Press.
Marcuse, Peter. (1975) "Residential alienation, home ownership and the limits of shelter policy". *Journal of Sociology and Social Welfare* 3: 181–203.
Netto, G. (2011) "Identity negotiation, pathways to housing and 'place': The experience of refugees in Glasgow". *Housing, Theory and Society* 28(2): 123–43.
Oxfam. (2018) "Hong Kong inequality report".
Planning Department (2018) "Land Utilization in Hong Kong 2018". Hong Kong SAR Government. Available at: www.pland.gov.hk/pland_en/info_serv/statistic/landu.html (accessed 27 August 2020).
Pollay, Richard W. (1987) "On the value of reflections on the values in the 'distorted mirror'". *Journal of Marketing*, 51 (2), 104–9.
Rapoport, A. (1995) "A critical look at the concept 'home'". In: Benjamin, D. N., Stea, D. and Saile D. (eds) *The Home: Words, Interpretations, Meanings and Environments*. Aldershot: Ashgate, 25–53.
Ryan, Richard M., Netta Weinstein, Jessey Bernstein, Krik Warren Brown, Louis Mistretta, and Marylene Gagné. (2010) "Vitalizing effects of being outdoors and in nature". *Journal of Environmental Psychology* 30: 159–68.
Saunders P. (1990) *A Nation of Home Owners*. London, Unwin Hyman.
Saunders, Peter and Peter Williams. (1988) "The constitution of the home: Towards a research agenda". *Housing Studies* 3 (2): 81–93.
Simmel, George. (1997) *Simmel on Culture: Selected Writings* (Ed. by Frisby, D. and Featherstone, M.). London: Sage.
Thurlow, Crispin and Adam Jaworski. (2010) "Silence is Golden: The 'Anti-communicational' Linguascaping of Super-elite Mobility". In: Jaworski, A. and Thurlow, C. (eds) *Semiotic Landscapes: Language, Image, Space*. London; New York: Continuum International Pub. Group, 187–218.

Tian, Yuhong and C. Y. Jim. (2012) "Challenges and strategies for greening the compact city of Hong Kong". *Journal of Urban Planning and Development* June: 101–9.

Transport and Housing Bureau. (2019) "Housing in figures 2019" Aug 30, 2019. Hong Kong SAR Government. Available at: www.thb.gov.hk/eng/psp/publications/housing/HIF2019.pdf (accessed 27 August 2020)

Wells, Nancy. (2000) "At home with nature: Effects of "greenness" on children's cognitive functioning". *Environment and Behavior*, 32(6): 775–95.

White, Mathew P., Ian Alcock, Benedict W. Wheeler, and Michael H. Depledge. (2013) "Would you be happier living in a greener urban area? A fixed-effects analysis of panel data". *Psychological Science* 24: 920–28.

8 A Conceptual Framework for Integrated Immersive Learning with Metaverse and Student-generated Media

Wong Pui Yun, Wong Wai Chung, and Shen Jiandong

Introduction

During COVID-19, disruptions to contemporary classrooms resulted in teachers having to adopt online learning, which also affects how students engage with one another. As classes start going online, one thing is clear: A new model can emerge on how teaching and learning can be conducted in a virtual space. While teaching can be directly facilitated with video conferencing tools, opportunities in other areas like engagement and assessments are lacking. This is especially so with reports of video conferencing fatigue and high stress with online teaching and learning (Nadler 2020). Despite this, one study (Butarbutar et al. 2021) reported an increase in digital literacy, which can help facilitate the adoption of technologies in the areas of engagement and assessments. For example, methods like social network student-generated media as well as more nascent communication technology like the metaverse can be explored further. Both topics can be regarded as somewhat underexplored, with little empirical evidence. The lack of understanding of these topics, especially when integrated, has motivated this study for which a conceptual model for online to offline, offline to online, and metaverse to the classroom is explored.

Social networks are important to students, especially in the 21st century. This is especially so when twenty-first-century skills like collaboration and critical thinking are essential for sustainable development (OECD 2019). Social interactions help students to practice these skills and can build contextual information from diverse perspectives. As conversations occur, if done correctly to exchange ideas, they can help to develop higher-order thinking in students. Given the rapid technological advances that occurred in recent decades, peer interactions are increasingly taking place online via various social media platforms. For example, having students take part in an online open call for problem solving can encourage crowdsourcing (Brabham 2013). However, social media technology may not always give positive student engagement or learning outcomes (Akcaoglu and Lee 2018; de Lima et al. 2019). For example, students may not always participate because they may not perceive these tools to provide a good sense of purpose. To enhance sense of purpose, it is useful to consider how social learning can take place on social media. It is possible to go back to the underlying theory, social learning theory, and how dialogic interactions can help to utilize observations and reflections in order to induce learning (Bandura 1963; Teo 2019).

Some studies use educational games to create engagement, and possibly a sense of purpose. Some studies with game design provide a framework where resources, objectives, and even social learning can be integrated into the curriculum (Mochizuki et al. 2021). Educational games can also create healthy competition to motivate student learning. For example, students frequently compare themselves to classmates in group learning environments. The competitive game design allows them to do so but additionally provides a space for them to compete and strive to do better than one another. Social game design can also allow for collaborative or cooperative learning (Lan 2020). For example, students studying a language require conversational practice with others to improve their skills. In a virtual environment, especially when using avatars, negative effects like embarrassment or fear of failure can further help to encourage participation. This is also known as psychological safety, which allows students to learn with the trust of interpersonal safety or fear of retribution (Edmondson 1999).

More recently, studies have shown that immersive and authentic learning experiences can be built with augmented reality or virtual reality to encourage motivation and participation. For example, in a study by Berg and Steinsbekk (2021), a VR simulation was used to allow students to practice the ABCDE approach used in a clinical patient assessment. However, studies have reported adverse effects of using highly immersive VR experiences, possibly due to dizziness, fatigue, and other side effects. Further, these technologies, especially when done in a highly immersive way, may not be scalable in terms of cost. More recently, interest in the metaverse has increased. Kye et al. (2022) describe the metaverse as a virtual space for participants to create and expand space–time together. According to Hwang and Chien (2022), the metaverse is a partially or fully virtual world in which the only rules that constrict the environment and its users are defined and limited by the creator's imagination. More importantly, Hwang and Chien (2022) are of the opinion that the use of non-player characters (NPC) can provide intelligent learning experiences. Dionisio et al. (2013) also describe metaverse architecture based on scalability with multiscene integration and NPC scripting. In this aspect, metaverse learning may be a viable method of student learning.

However, it is unclear how a metaverse can be integrated with social-media-style interactions, particularly where student-generated content like pictures, videos, and dialogue can occur. Some studies suggest scenarios of how the metaverse can be used in education. For example, Hwang and Chien (2022) recommended some scenarios where the metaverse may be relevant. One includes that the replication of a real-life learning environment may be

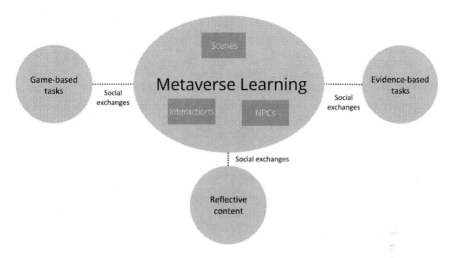

Figure 8.1 Social-integrated metaverse learning conceptual model.

difficult to simulate in the classroom. Providing evidence of real-life activities, including reflecting on them, is also important. Another example is when varying scenarios and experiences are necessary. In a contemporary classroom, creating multiple characters or stakeholders within an ecosystem may be a challenge to replicate and monitor. Using game characters in a metaverse may help to facilitate, especially where social learning is involved. One study by Mochizuki et al. (2021) showed how incorporating peer-to-peer interactions in an educational board game can create interest and motivation. However, evidence of online game-based learning with social learning is lesser known. As social interactions and varying experiences are a significant attribute of metaverses (Dionisio et al. 2013), it is important to understand whether these methods are well-received even in a traditional social media environment. After which, recommendations on integrating them on a metaverse could then be considered.

Thus the purpose of this study is to propose a conceptual model for social integrated game-based metaverse learning. Aspects of the model are tested, and a possible integration are discussed. See Figure 8.1.

Method

Participants

This study involved four groups of participants. Group one was six students recruited to test and evaluate a purpose-built education metaverse-learning environment. Group two consisted of 23 students taking part in an economics course utilizing student-generated content on a private social media application. Group three consisted of nine students taking part in a data science course utilizing student-generated content on a private social media application. Group four consisted of 32 students taking a sustainability course, leveraging a private social media application. Participants were all taking undergraduate degrees in a liberal arts institution in Hong Kong and mostly resided in Hong Kong during the study. Participants utilized their own mobile phones for the study and had a relatively good Internet connection.

It is anticipated that comparing and integrating cross-disciplinary courses into a conceptual model can widen the scope of the conceptual model and increase impact.

Assessments and Measures

In the proposed conceptual model, the underlying technology used to facilitate engagements is important. It can be said that both social media and metaverse technology may provide features that support interactions and tasks. While it is possible for content like presentations and videos to be uploaded on social media and shared with students, content is pretty two-dimensional and often lacks two-way interaction. During COVID-19, reports of video fatigue also mean there is a lower chance of participation. Thus, for the purpose of this study, the metaverse is considered the main interaction and engagement medium. The other concepts utilized for the conceptual model for this study are collective efficacy (Bandura 2000; Glassman et al. 2021), authentic learning (Farrell 2020), ubiquitous learning, and game design (Mochizuki et al. 2021). The way in which these frameworks and pedagogies were adapted is represented in Table 8.1. The resulting conceptual model is represented in Table 8.2.

The activities and tasks for the study were different and used different pedagogical designs. Tasks on social media can be integrated for completing activities, some of which can require offline activities. For example, to capture evidence of completing offline activities, like waste auditing in group four, videos can be captured offline and put together into a presentation to be shared. Comments can then be uploaded for idea exchange. Reflective videos can also be used to document learning experiences, which was used in group three. Game-based design can also be used to create simulations, for housing transactions, as used in group two. The overall task design and data collection are described in Table 8.2.

The overall conceptual model for this study is represented in Table 8.2. The properties of the metaverse, primarily the use of integrated and multiple scenes, social interactions, and the use of non-player characters (NPC), would form the primary base of participant communication. While related activities like game-based learning, evidence-based tasks, and reflective learning were conducted separately in this study, they are designed to be complementary to the metaverse. According to Toh and Kirschner (2020), some factors that promote self-directed learning in video games are the release of new information

Table 8.1 Conceptual Framework with Metaverse and Social Learning

Theory	What Is Relevant	How It Can Be Adapted
Collective efficacy: Bandura (2000) and Glassman et al. (2021)	Participants in the learning system contribute both in an individual and group capacity. In particular, social learning theory describes that participants learn through observation, within the environment	The metaverse will form the centre of activity, in a multiplayer level where students can engage, and participate in activities, particularly with non-player characters as well as other players. The application used for the study is Classlet by Soqqle (https://soqqle.com). Classlet is a purpose-built metaverse learning application that is designed for plug-and-play scenes.
Authentic assessments: Farrell (2020)	Authentic learning goes beyond learning in a realistic environment but also psychological factors like challenges, varying experiences, and collaboration	Activities are designed around challenges and varying experiences with peer-to-peer interactions designed in multiple scenes.
Ubiquitous (e.g. mobile learning) and game design: Churchill (2016) and Mochizuki (2021)	The learning framework requires considerations of resources, support, evaluation, rules, roles, and planning for social interactions where possible	Tasks in the metaverse or social media are designed with rules around a game system to promote interactions and engagement. In this study, a video application, Soqqle video-app (https://soqqle.com) was used. The mobile application supports video uploading and peer-to-peer comments.

over time, conducted through game-based challenges; scaffolding, which will be supported by reflections and evidence-based activities; and a safe space to experiment. As a pilot study, feedback on these complementary tasks will be utilized to propose a method that can be integrated into the metaverse.

Results

Metaverse learning: Participants were able to complete the tasks on the scene successfully, although some issue bugs, like error messages in the scenes,

Table 8.2 Groups and Scope of Study

Group	Topic and Period of Study	Technology Used	Participants	Tasks That Participants Completed	Data Collection and Analysis Method
1	Sustainability April 2022	Metaverse	6	Interact with non-player characters to review 9 content and complete multiple-choice questions in 20 minutes	Questionnaire using adapted technology adoption model and authentic learning
2	Economics March 2021	Social media	23	Participate in a game where participants role-play as buyers and sellers of property. Transactions are made using comments.	Self-assessment in form of essays, with data analysis using an authentic learning framework
3	Data science May 2022	Social media	9	Prepare up to 10-minute reflection videos and upload for peer-to-peer commenting	Correlational analysis using duration of peer video watching and peer comments against video scores
4	Social media Aug 2021	Social media	32	Prepare up to 10-minute videos to evidence daily waste reduction and upload for peer-to-peer commenting	Correlational analysis using duration of peer video watching and peer comments against final scores

were reported. Screenshots of the NPCs and quest dialogs are shown in Figure 8.2. Some participants in group one shared that they were familiar with the metaverse environment, while those who attempted it the first time only required 3–4 minutes to adapt. It was also reported that the realistic environment encouraged participation and that new ideas were shared about creating tasks in the scenes linked to missions. Here are some illustrative quotes:

A Conceptual Framework 119

Searching for quest NPCS

Starting new quests

Reviewing text and completing tasks

Figure 8.2 Screenshots of the metaverse learning application, Classlet.

I do play some games like this. I played Apex, Decentraland, or Roblox since the metaverse topic is getting popular. And I just want to know what it is about. Apex is more about the environment detail and texture. But the Decentraland or Roblox will be focusing more on the interactive and the reality inside the game. And these are new to me.

I think the wide range of content will make me explore it more. From the daily habit to complete some missions. But I think something more related to real-life something I can't do in reality, I will be more interested in it. Also, I think the Q&A section can be more simplified, sometimes it will make me less patient when I have to deal with long text repeatedly.

Since it was my first gamified learning experience, it took me around 3–4 minutes to adapt to such an environment and get the hang of the way I need to move around and interact with other characters.

Certainly, the realistic environment does encourage me to be proactive in the learning process by keeping me curious about what scenery is going to be next and what type of tasks and people I am about to meet. Especially, after a long period of studying only on learning platforms and via online conference tools, the gamified learning in a real school environment made me more interested in the learning process.

Yes, I think especially younger students will find game-like challenges interesting. The shorter attention spans of kids, which is natural, and even of adults due to the abundance of information on the Internet makes game-like challenges a great solution for both keeping a person's focus and facilitating his learning process.

120 *Wong Pui Yun, Wong Wai Chung, and Shen Jiandong*

Figure 8.3 Screenshots of Soqqle video application with game-based learning.

Game-based tasks: Thirty-five properties were posted on the Soqqle video app, for which 60% were sold (21). See Figure 8.3. The coding of the reflective essays showed that reflection and knowledge transfer were the common themes. Some highlighted realism and challenge within the process, which are factors for authentic learning assessments (Farrell, 2020). Illustrative quotes are shown here:

> I have a strong insight after selling property to these two diverse buyers, which makes me understand the importance of having savings.
>
> In the last trade round, 9 and 13 were still available for purchase. Thus, A4006 was put into the market as planned. Unfortunately, it didn't sell out.
>
> In addition, the house information I posted on the Soqqle app also specifically stated that my house is suitable for Buyer No. 9 and Buyer No. 13, and I hope they can see my house as soon as possible. Although they did not buy my house in the end, I have tried my best.

Reflective content: Nine videos were posted on the Soqqle video app, for which 16 comments were shared. See Figure 8.4. Out of the 16 comments, five were elaborate and provided substantial feedback about data science methods. Illustrative quotes:

> Excellent explanation of Model, Loss Function, Optimizer and Hyperparameters and point out the limitation and future work clearly.

A Conceptual Framework 121

Figure 8.4 Screenshots of Soqqle video application with reflective learning.

Moreover, present with appropriate time allocation and pace allocate time appropriately and manage time effectively. Appropriate pictures that display on the slide to show the result and show balanced posture, enthusiasm and confidence.

"Your model is good in that you can run for both cross-entropy loss and also MSE loss. For the design and functions part, you have used the softmax regression to find the loss function by using different variables that fulfil all the requirements of the project. Also, for the Demonstration and Performance part, you are able to finish those five questions with the reasonable answer as well as the graph. Lastly, for the conclusion part, you have provided some limitations of the in-use of CNN and the future views for CNN and also a better model in the future to increase the accuracy.

Evidence-based tasks: Sixty-two waste audit videos were posted on Soqqle video-app, with 139 comments exchanged. See Figure 8.5. Focus groups suggest that the method largely improved idea generation, enjoyment, and experience. Students felt that watching the video content of peers helped develop their ideas. The format of videos is effective as the realism of the format helps enhance understanding of the content and association with ecoliteracy.

Through watching classmates' household waste audit videos, I learned a lot about green practices from my classmates.

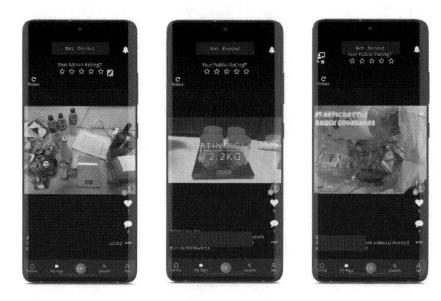

Figure 8.5 Screenshots of Soqqle video application with evidence-based learning.

After uploading my first week's recording video, I received comments from my classmates. The comment said I threw away the trash without removing the tape from the carton, which is not good for recycling.

Discussion

The results of the metaverse cases suggest that the 3D metaverse virtual environment is a suitable place for participants to browse to receive knowledge. Participants indicate that they feel comfortable and relaxed in engaging with objects in the virtual environment. Evidence of social media-based activities also suggests that there may be benefits in incorporating hybrid online/offline assessments. The adapted conceptual model is especially useful for topics like sustainability, which require students to understand issues closer to real-life scenarios. It is also possible to incorporate storytelling methods to deliver content in a more interactive and active manner. Participants recommended activities that allowed them to be placed directly into the scene instead of just playing like a third person. While the study did not particularly have related scenarios, metaverses can also be designed for multiplayer scenarios which can encourage interactions for social learning.

Participants concurred that the ability to review the content submitted by their peers was helpful, as it allowed them to develop their submissions.

A recent neuroscience theoretical study (Ramsey et al. 2021) has suggested that visual observations can create an internal reward mechanism that uses the knowledge gained from others, which explains why watching content from peers can help develop one's ideas. It's also interesting for students to see videos of peers and learn from them. For example, when watching content from peers, they could appreciate the pronunciation and grammar of others. Further, in real life, presentations could only be watched once, while the video format allows the content to be watched again and again. The majority of participants felt that the learning method promoted peer engagement, as the ability to receive likes and comments improved interaction, making the learning experience more productive and enjoyable.

Social learning can be integrated into games where students role-play as buyers and sellers of properties. The model that we used for RASE (resource, activity, support, evaluation) was a good fit when integrated with Mochizuki's game design model. Social elements in game design prove to be useful, but the amount needed remains uncertain. While competition can be a useful motivator for learning, it may also have the opposite effect, stressing and demotivating students (Xu et al. 2021). These discordant results on the effectiveness of competitiveness indicate the need for further research to obtain more insight into applying social elements to VLE. In addition, too much of a social game feel can cause students to lose sight of the main goal: education. A study on social networks for language learning suggests that students shift from writing practice to socialization with native speakers (Lyu et al. 2022). While this is effective for language learning, it can become a hindrance to other educational topics. For example, students socializing for the purpose of discussing climate change can easily shift off topic. Specific research into a conceptual framework to integrate game-based learning for VLE is required for a better understanding of social elements.

In this study, metaverse learning effectively combines an educational framework with a virtual 3D environment in which users interact so as to allow them to gain knowledge. Virtual reality studies have shown that immersive environments may produce adverse health effects such as nausea, headache, fatigue, dizziness, etc. (Moro et al. 2017). However, more modern concepts like a metaverse with playable avatars or open-world spaces can help reduce these effects. In addition to utilizing metaverse-based learning and the like, producing a conceptual framework with low fidelity and/or low immersion may be more cost-effective. Games do not always require high immersion to produce the desired outcomes. The outcomes can be even more substantial if integrated with methods like social learning. We share some examples of how the social methods used in this study can be integrated into the metaverse learning environment in Table 8.3.

Table 8.3 Recommendations of Social Learning Methods in Metaverse Environment

Pedagogy	Current Method	Proposed Metaverse Method
Game-based learning	Image-based housing simulation where students upload real-life pictures representing for-sale property	Virtual exhibition where participants can upload visual representations of housing sorted by housing locations. Shoppers can browse the exhibition and ask questions to NPC, with predefined dialogues.
Reflective learning	Video-based reflective videos to document learning experiences	Integrating direct and timely feedback through dialogic interactions in the metaverse as participants complete tasks through a storyline as a scaffolding method
Evidence-based learning	Video-based evidence videos to showcase learning output	As participants browse the virtual environment to receive information, they can upload real-life evidence to complete tasks or assessments. They can also browse the content of peers to give feedback.

Conclusion

This study is the first that the authors could find that incorporates a metaverse pedagogical conceptual model for education integrated with social-learning-related methods. For example, the study evaluated how game-based learning, evidence-based learning, and reflective learning could be integrated into social activities in the metaverse. Other social methods could also be incorporated, based on peer-to-peer and observational learning, to expand the effectiveness of the metaverse in a learning context. The opportunity is significant due to the possibility to increase collective intelligence in an education environment. This is especially important when a game-based metaverse learning environment provides an alternative to the video methods commonly used due to COVID-19. The conceptual model of this study can be beneficial for practitioners exploring new innovative online methods like the metaverse.

References

Akcaoglu, Mete and Eunbae Lee (2018) "Using Facebook Groups to Support Social Presence in Online Learning", *Distance Education* 39(3): 334–52, available at https://doi.org/10.1080/01587919.2018.1476842

Bandura, Albert (2000) "Exercise of Human Agency through Collective Efficacy", *Current Directions in Psychological Science* 9(3): 75–78, doi: 10.1111/1467-8721. 00064.

Berg, Helen and Aslak Steinsbekk (2021) "The Effect of Self-practicing Systematic Clinical Observations in a Multiplayer, Immersive, Interactive Virtual Reality Application versus Physical Equipment: A Randomized Controlled Trial", *Advances in Health Sciences Education* 26(2): 667–82, available at https://doi.org/10.1007/s10459-020-10019-6

Butarbutar, Ranta, Angla F. Sauhenda, Marnina, Seli Marlina Radja Leba, Hanova, and Wahyuniar (2021) "Challenges and Opportunities of Accelerated Digital Literacy during the COVID-19 Pandemic", *Hong Kong Journal of Social Sciences* 57: 161–68.

de Lima, Dhanielly P.R., Marco A. Gerosa, Tayana U. Conte, and Jose Francisco de M. Netto (2019) "What to Expect, and How to Improve Online Discussion Forums: the Instructors' Perspective", *Journal of Internet Services and Applications* 10(1): 22, available at https://doi.org/10.1186/s13174-019-0120-0

Dionisio, John David N., William G. Burns III, and Richard Gilbert (2013) "3D Virtual Worlds and the Metaverse: Current Status and Future Possibilities", *ACM Computing Surveys* 45(3): 1–38, available at https://doi.org/10.1145/2480741.2480751

Edmondson, Amy (1999) "Psychological Safety and Learning Behavior in Work Teams", *Administrative Science Quarterly* 44(2): 350–83, available at https://doi.org/10.2307/2666999

Farrell, Carlyle (2020) "Do International Marketing Simulations Provide an Authentic Assessment of Learning? A Student Perspective", *The International Journal of Management Education* 18(1): 1–13, available at https://doi.org/10.1016/j.ijme.2020.100362

Glassman, Michael, Irina Kuznetcova, Joshua Peri, and Yunhwan Kim (2021) "Cohesion, Collaboration and the Struggle of Creating Online Learning Communities: Development and Validation of an Online Collective Efficacy Scale", *Computers and Education Open* 2, 100031, available at https://doi.org/10.1016/j.caeo.2021.100031

Hwang, Gwo-Jen and Shu-Yun Chien (2022a) "Definition, Roles, and Potential Research Issues of the Metaverse in Education: An Artificial Intelligence Perspective", *Computers and Education: Artificial Intelligence* 3, 100082, available at https://doi.org/10.1016/j.caeai.2022.100082

Kye, Bokyung, Nara Han, Eunji Kim, Yeonjeong Park, and Soyoung Jo (2021) "Educational Applications of Metaverse: Possibilities and Limitations", *Journal of Educational Evaluation for Health Professions*, 18, 32, available at https://doi.org/10.3352/jeehp.2021.18.32

Lan, Yu-Ju (2020) "Immersion, Interaction and Experience-oriented Learning: Bringing Virtual Reality into FL Learning", *Language Learning & Technology* 24(1): 1–15, available at http://hdl.handle.net/10125/44704

Lyu, Boning and Chun Lai (2022) "Learners' Engagement on a Social Networking Platform: An Ecological Analysis", *Language Learning and Technology* 26(1): 1–22. https://doi.org/10125/73468

Mochizuki, Junko, Piotr Magnuszewski, Michal Pajak, Karolina Krolikowska, Lu Kasz Jarzabek, and Michalina Kulakowska (2021) "Simulation Games as a Catalyst for Social Learning: The Case of the Water-food-energy Nexus Game", *Global Environmental Change*, 66, 102204, available at https://doi.org/10.1016/j.gloenvcha.2020.102204

Nadler, Robby (2020) "Understanding 'Zoom fatigue': Theorizing Spatial Dynamics as Third Skins in Computer-mediated Communication", *Computers and Composition* 58, 102613, available at https://doi.org/10.1016/j.compcom.2020.102613

OECD (2019) "OECD Future of Education and Skills 2030 Concept Notes", available at www.oecd.org/education/2030-project/teaching-and-learning/learning/core-foundations/Core_Foundations_for_2030_concept_note.pdf

Ramsey, Richard, David M. Kaplan, and Emily S. Cross (2021) "Watch and Learn: The Cognitive Neuroscience of Learning from Others' Actions", *Trends in Neurosciences* 44(6): 478–91, available at https://doi.org/10.1016/j.tins.2021.01.007

Toh, Weimin and David Kirschner (2020) "Self-directed Learning in Video Games, Affordances and Pedagogical Implications for Teaching and Learning", *Computers and Education* 154, 103912, available at https://doi.org/10.1016/j.compedu.2020.103912

9 Problems of Exacerbation to Dasein in the Modern Technological World by Use of the Early Heidegger's Theories
Readiness-to-hand and Presence-at-hand

Lau Hok-Yin

Introduction

In this era of information technology, in our everyday lives, we can by no means avoid our critical connections with the use of technological appliances and gadgets in various activities ranging from payments, learning and teaching, to working and interpersonal communication. We are into this modern lifestyle, by which, first, we are subconsciously using technology in substitution for our primitive laborious lives, and, second, this subconscious act is what today's people cannot live without. The reason for the utmost importance and the expansion of technological lifestyles has something to do with expansion of people's living network from domestic regions to the global contexts in which people do not need to put much effort, say money and time, in reaching. These technological essentials in relation to our modern lives are taken for granted and require agreements across the globe. This issue is also what Heidegger mentions especially in his later years. Usually, when it comes to technology, it is the late Heidegger taking on the role for thematic investigation. However, does it mean that the early Heidegger's ideas are not suitable for an investigation of the issues of technology related to Dasein? No. As a matter of fact, based on what Heidegger always emphasizes in *Being and Time*, Being-in-the-world is essentially the most basic existential mode of Being of Dasein, which implies that I myself, as Dasein in any of the modes of Being, am invariably connected to the world by disclosing itself, one kind of which is a technological world as chosen for this thesis, for as long as I am living Dasein, dwelling in it. The Being, Being-in-the-world, lays a solid foundation for different possibilities of and other modes of Being of Dasein, given its openness to the world, including readiness-to-hand and the secondary (founded) but deficient mode of Being, present-at-hand understanding of the world. While technology serves as assistance to us in wide-ranging aspects of our lives, we invariably deal with it to solve problems involving other people—the establishment of the web of essential Dasein's

DOI: 10.4324/9781003376491-10

social practices in Being-in-the-world and its everyday readiness-to-hand, "in-order-to" and "for-the-sake-of-which", or we theorize it for the exploration of its physical occurrences—the Subject–Object dichotomy which is inappropriate to Dasein's primordial existential modes (Schmidt 2006: 51). Without doubt, the early Heidegger's theories do provide us with relevant ideas for this paper. Presence-at-hand is one of the problems that Dasein may face in the technological world. However, in this paper, more of a problem should be the *exacerbation* of the problems, as this paper would argue that the technological world would exacerbate the exposure of the presence-at-hand. All in all, this paper would mainly focus on possible problems posed to Dasein by the *exacerbation* resulting from technology by leveraging the early Heidegger's notions of presence-at-hand versus readiness-to-hand. The research questions addressing the *problems* are:

1. How does the technological world trigger Dasein's present-at-hand understanding of the physical occurrences of technology more easily than the primitive life world?
2. Why is question 1 in opposition to the basic existential essence of Dasein?

Introduction to the Basic Existential Mode of Being, Being-in-the-world, and Being-in-the-technological-world

Being-in-the-world, Understanding and Possibilities

To Heidegger, before answering any questions of phenomenological realities outside of the Being of human beings, there is a need to take the initiative to conduct investigation of human beings' own facticity—Dasein's particular mode of being under a particular context[1] (Schmidt 2006: 52). That is to say, the inquiry into the existential essence of human beings' Being—why am I here, but not inexistent?—is the more basic and fundamental inquiry into the nature of existence from human beings' "realm of average everyday experience"[2] (Anowai and Chukwujekwu 2019: 1; Lawhead 2002: 536). This kind of inquiry is thus regarded as a task of fundamental ontology by reflecting on our everyday life experience where we are immensely into our daily lives—absorbed in our daily routines as what we have normally done without doubt and suspicion, as are "walking to school" by students, "greeting superiors" by employees, and "cooking food" by cooks.

Human beings are regarded as entities whose Being is characterized as Dasein,[3] i.e. "Being there" in our literal interpretation of the wording. The reasons for naming human beings' Being as "Being there" are manifold. Firstly, human beings are invariably situated and living in the world one way or another, and, secondly, Being there concerns *how* one is being there (Schmidt 2006: 53). The way that human beings are inextricably "embedded" in the world is diverse, diverse in the sense that mankind's existential way of living accommodates a set of Dasein's own-most possibilities of being itself projected to the future[4] (Schmidt 2006: 55). A particular set of possibilities, be they as very subtle as a change of a happy smile or as very conspicuous as an obviously aggressive behaviour, can be interpreted as a set of *choices* as to *how it will be*, made by Dasein based on its interpretation of its own current situation with its own self-understanding (Schmidt 2006: 56, 63). Moreover, this *selective* activity is, proximally and for the most part, conduced subliminally by Dasein when living in the average everyday world at every moment—a world full of the Others, the They, that affect Dasein's decision making. This is one point. Another point to notice is that the act of dealing with a set of possibilities must be realized and embedded through the basic existential mode—Being-in-the-world—because a human being cannot *select* a possibility to his/her own future alone without situating him-/herself in the world where s/he is essentially and invariably responding to and dealing with surroundings; that is to say, every possibility selected is a possibility in response to (at least) a matter from the world, be it a party of people/person, an inanimate object, or an incident.

The reasons for these essential features of (1) Being-in-the-world as one of the existential modes of Dasein and (2) the activities of making a choice from possibilities have something to do with human beings' states of mind such that Being-there carries (primordial, very fundamental) understanding[5] as an existential structure to respond to the world. The details are as follows.

Understanding as a Characteristic of Being-in-the-world

> *Understanding of being is itself a definite characteristic of Dasein's Being.*
> (Heidegger 1962: 34/12)

As Dasein as Being-in-the-world, we, at every moment until we die with no consciousness, have our states-of-mind situated in the world, which brings in affectedness, due to our everyday (primordial) understanding of the world: The moment Dasein understands, it finds itself existing in the world where it understands itself and understands that the world matters, as a *mooded* being (Elpidorou and Freeman 2015: 661). The way Dasein carries the incessant understanding activity as an existential structure of our state of mind is the way Dasein is being emotionally affected by the world one way or another (ibid.: 664). Dasein understands itself in the relation to the world-hood, which affects Dasein by entering and changing Dasein's perception/affectedness and Dasein-world interconnectedness, by which to indicate Dasein's ontological constitution and its prior embeddedness in the world (ibid.: 668).

Openness, a Mutual Relationship between Possibilities and Understanding

> *In hermeneutics, what is developed for Dasein is a possibility of its becoming and being for itself in the manner of an understanding of itself.*
> (Heidegger 1999: 11)

Other than the features of our primordial everyday understanding just described, Stefani and Cruz (2019) mention that our everyday understanding activities are invariably "a non-stopping flow of understanding". That is to say, the moment Dasein understands, it is ready to select a possibility for the future. To put it in the context of recurring understanding, the understanding activities, along with the selection of possibilities towards the future, are on the run constantly so that, at every moment, Dasein is becoming and being for itself repeatedly and endlessly until it dies (Stefani and Cruz 2019). Furthermore, the relationship between understanding and possibilities can be interpreted as *openness*. It is not about the opening of a concrete object but about an introduction of *something* new, say a new decision made about a set of possibilities or a new understanding about the selection of a particular possibility in Dasein's choice. Therefore, I argue that openness operates in a mutual bidirectional way that Dasein projects itself in the world in temporality (Stefani and Cruz 2019). For instance, on the one hand, my current understanding of a fire in the my kitchen (as the fire in the kitchen manifests itself to me now) becomes the openness to a set of possibilities, with one that I step forward to put the fire out rather than escape from it; therefore, understanding becomes "*openness to possibilities as a constitutive movement within time*" (Stefani and Cruz 2019). On the other hand, my decision to *put out the fire* becomes the constitutive openness of my tightly subsequent understanding—"Oh, the fire is out now, and the kitchen is safe now!"—that is, something has happened and has affected me, and then I perceive myself as part of the past tradition in my understanding of this ongoing moment. Each turn between understanding and the selection of possibilities affects me as Dasein projects itself in the world towards the future.

Being-in-the-Technological-World

Dasein with diverse possible existential modes of Being is itself living Being-in-the-world. However, besides what Heidegger proposes as Dasein's existential modes of Being such as Being-with-Others, given that there is always a set of many possibilities to characterize Dasein's existence, human beings have made full use of them to project themselves to the future by living ordinarily in this project-oriented and equipmental world (O' Brien 2014: 537).

The transformation of human beings' ways of living has illustrated two important points: First, when it comes to the transformation of human beings' ways of living, I refer to the transformation in the collective sense, which embodies the advancements of human civilization; second, based on the first point, the fact embodies the existential essence of Being-in-the-world where Dasein cannot operate without openness to the world of surroundings, including other people and inanimate objects, without which the advancements of human civilization cannot happen.

In the course of human civilization, we have brought in technology and have integrated it into our everyday lives and have kept advancing it. The way we integrate technology into our everyday lives comes along with how we take it for granted. This "take-it-for-granted" approach has made us accommodate technology as a part of our lives such that we cannot live with it. Some may argue that today's people can "survive" without technology. This argument is valid, but it only reflects a tip of the iceberg. While this paper focuses on the modern technological world, we view it from the global perspective in the general sense. Today's technological applications are different from what Heidegger might have conceived of. While Heidegger is opposed to them, today's people take them for granted.

On a daily basis, we make full use of our phones to text people, make transactions, enjoy entertainment, search information, and so on, besides taking transports to destinations to replace walking. Apart from them, we use appliances to do our homework and chores in order to reduce our physical bodily movements, and we make acquaintances online, albeit virtual, instead of doing so in face-to-face settings. Even though we leverage technology to live our everyday lives, we invariably feel connected to the world; such an experience can be made known from a case that when a pair of lovers see each other's messages popping up via WhatsApp, they still sense the interconnectedness. This way of living with a kind of interconnectedness also embodies human beings' fundamental existential essence, Being-in-the-world. However, as human beings are living their lives with technology as a medium and as a part of their lives, Being-in-the-world can be viewed as Being-in-the-technological-world in today's context.

Technological World, Present-at-Hand Understanding and Ready-to-Hand Understanding

Technological World and Ready-to-Hand Understanding in the Primordial Context

In Dasein's average everydayness, we, as Dasein, are subliminally engaged in the employment of things as the ready-to-hand[6] situated in the pragmatic world. This is a pragmatic sense in which Dasein really knows that the way of using a thing is meaningful in one particular context rather than another[7] because the thing shows itself to Dasein as a useful thing in the particular context (Schmidt 2006: 64). That is, Dasein becomes absorbed in the ready-to-hand circumstances of a totality of things showing themselves to Dasein. However, when considering the world as a technological world, I argue that our absorbed-ness becomes more intense. One of the reasons for this has something to do with the advantages that the modern technological life brings us compared with a world without technology and with less technology than we have now. Furthermore, the more intense absorbed-ness in the way of technological life is due to obsession. "Obsession" is so close to the indulgence of children in game playing that human beings find it hard to get rid of technology in their lives because they are addicted to it. Taking the consequence of obsession into account, we may find that people of today are more likely to be under tremendous stress and experience anxiety when deprived of what they have taken for granted—accustomed reliance on advantages brought on by technology. However, as I argue, even though the degree of obsession with technology is greater in today's contexts, it does not mean that Dasein's exposure to technology as present-at-hand is reduced. Instead, it is quite the contrary. Greater obsession with technology may result from Dasein's consideration of the advantages and convenience brought by technology, while greater exposure to presence-at-hand may come from greater frequency of various problems of technology and technology itself.

Manner of Dasein's Use of (Technological) Things as Ready-to-hand

In the modern technological world, Dasein always uses the surrounding technological items as ready-to-hand without viewing them from a scientific and philosophical perspective. To describe the ready-to-hand understanding of technological items more vividly and in our everyday manner, we can describe them so naturally and subliminally that even without looking at their structures and positions, we can somehow accurately locate them and pick them up for uses in certain ways to fulfil our intended needs but not mull over them consciously. For instance, when I wake up at 6:00 in the morning, from the moment I step out of my bed, the horizon of my surrounding is still unclear due to feeling sleepy (supposedly until I wash my face to wake myself up). However, I can somehow go to the toilet accurately without stumbling against any objects, successfully pick up my toothbrush, and put it into my mouth.

To put it in the context of technology, when I see message notifications of WhatsApp pop up on the screen of my phone on the desk in front of me, I respond to this stimulus by naturally picking up the phone without paying attention to my hand position and movement, pressing my thumb on the sensor without looking at the position of my thumb, touching the icon of WhatsApp on the screen without entirely focusing on the contact of my finger with the icon, and finally typing words from the given keyboards on the screen without wholly eyeing each of the letters; the manner of their use is attested as follows

> *The less we just stare at the thing called hammer and the more actively we use it, the more original our relation to it becomes and the more undisguisedly it is encountered as what it is, as a useful thing.*
>
> (Heidegger, 1996: 69)

The way human beings employ technological appliances, gadgets, and equipment to bring their functions into full play is the same as the way we deal with non-technological Things. This description of my conscious state of my using things is just one of the features of the ready-to-hand understanding.

Referencing Characteristics of (Technological) Things as Ready-to-hand

In addition, the significant framework of referential totality of (technological) Things is another feature of the ready-to-hand understanding in this equipmentally oriented world (O'Brien, 2014: 537).[8] What this means is that when using or manipulating a thing, it is usually not employed alone. Instead, Dasein uses a thing in a framework of relations to other things; that is, the use of a thing leads Dasein to use another thing such that the former use motivates the emergence of the latter use in the sense of interrelatedness and of course "in a ready-to-hand environment context of equipment ..." (Heidegger 1962: 154). This is the mode of "in-order-to-which"; that is, Dasein does one thing with a thing in order to do another thing with another thing—the structure of "a reference of something to something" in the relational sense (Schmidt 2006: 64). This is not a simple two-step process but a recurring process that possibly involves more than two things in their frame of referential totality. The use of things in the referential totality by which the mode of "in-order-to-which" is implied is ultimately meant to fulfil a purpose (or more) of Dasein's using things—the mode of "for-the-sake-of-which" that refers to what Dasein wants to achieve/accomplish by the act (Schmidt 2006: 65, 66). To describe the modes of "in-order-to-which" and "for-the-sake-of-which" of employing things in the context of technology, to commence with the use of emails is to provide us with a sound example. In Dasein's average everydayness in technological contexts, given that its purpose is to fetch a name list from another Dasein (the colleague) ("for-the-sake-of-which"), it uses a computer to send an email to its colleague to ask him/her to download a name list from a school portal; what follows is that the colleague receives the email on his/her computer, downloads the name list requested, and orders the printing job to print the searched name list out before handing it to Dasein (me) (a series of "in-order-to-which"). It is important to note that the frame of reference of totality of the uses of technological things is disclosed both sequentially and simultaneously. The former, *sequentially*, can be referred to as a one-to-the-next-one process—the first thing is used as a necessary condition without which the second thing cannot be naturally used, while the latter, *simultaneously*, can be understood as a synchronous use of (at least) two things to fulfil my purpose.

Problems of Present-at-hand Understanding in Opposition to Ready-to-hand Understanding

Technical Problems of Technology, Non-technical Problems of Technology and Problems to Dasein's Being-There Brought on by Technology

Though able to use technological things in its average everydayness as if Dasein uses non-technological Things, Dasein is very likely to encounter many problems with technology, be they expected or unexpected. Taking *problems* of related to technology into consideration, we shall differentiate *technical problems of technology* from *non-technical problems of technology and problems* and from *Problems to Dasein's Being-there Brought on by Technology*. Both of them are different from each other, but it does not mean that they do not share a point of interface.

Problems to Dasein's Being-There Brought on by Technology

The latter, *problems to Dasein brought on by technology*, refers to problems brought on by technology posed to the existential essence of Dasein's basic existence, Being-there (Being-in-the-world). The problems posed to Dasein do not mean the disappearance of Dasein's existential mode, Being-in-the-world. Instead, they are affecting Dasein's primordial modes of existence, such as the exacerbation of the emergence of presence-at-hand in the use of things and the exacerbation of inauthenticity.

Technical Problems of Technology

The former, *technical problems of technology*, refers to the technical problems of technological gadgets, appliances, and equipment that we use, such as the sudden breakdown of a computer and various kinds of errors. These kinds of problems can be expected and unexpected. If the problems of technology are expected, then perhaps the problems themselves do not matter to Dasein because the problems *have been* predicted and can be handled at least psychologically. On the contrary, if the problems of technology are unexpected, then they themselves matter to Dasein, which implies the need of Dasein's reflective thinking about and repeated focus on the technical problems of the technology and the related physical structures and details. Furthermore, no matter whether the Dasein is experiencing an unexpected or expected problem, it is likely to focus on the technical problem itself from a scientific perspective and with a "thinking-twice" approach. The reason for this is that, no matter whether or not the technical problem is expected, Dasein still needs to solve it. The act of solving technical problems involves a scientific focus and reflective thinking. This thinking approach and perspective can possibly lead technical problems encountered by Dasein to the *problems to Dasein brought on by technology*.

Non-technical Problems of Technology

Another kind of problems related to technology may not be considered as problems generally but considered as normal *features* of technology. The features of this kind of *problems* are not the recognizable technical problems or errors. Instead, the features are about the structures and the modes of technology itself in the informationalised world (Christ 2015: 63). The structures and the modes are addressing and worrying Dasein itself without displaying any of the technical problems or errors as previously discussed. Therefore, this would be regarded as *non-technical problems of technology*. For instance, Christ (2015: 63) claims:

> *Both the things of reality and everyday phenomena are reduced to data and mathematical quantities (and processes to algorithms).*

In the informationalised world, the way we use technological devices, mostly computers, involves countless processes of "calculable quantities and relationships", figures, numerical units, statistical operations and so on, for the most part, in our daily lives (Christ 2015: 63). In this technological world, Dasein finds it normal to be exposed to the processes of items as they serve as instruments to meet Dasein's ends. Unlike *technical problems of technology*, which Dasein may consider negative and urgent to be solved at once for continuous usage, the non-technical problems of technology can be regarded as something neutral and maybe even positive from our ordinary and everyday perspective as the informationalised processes run faster, more accurately and helpfully. However, the perception of the faster, more accurate and helpful processes of the informationalised processes is Dasein's surface perception only as Dasein finds enjoyment in them all. If taken to a deeper reflection, it shows us that this informationalised world of technology may have great potential to make itself become the *existential problems to Dasein brought on by technology* (Christ 2015: 58–63).

Hypothesis (1): Relationships between Technical Problems of Technology, Non-technical Problems of Technology and Problems and Dasein's Being-There Brought on by Technology

The relationships among technical problems of technology, non-technical problems of technology, and Problems to Dasein's Being-there Brought by Technology are that:

Hypothesis (1): The former two problems related to technology are likely to become Problems to Dasein's Being-there Brought on by Technology, as indicated in the Figure 9.1. Such a hypothesis will be tested in the next section.

Justification of the Problems to Dasein's Existence Posed by the Problems Related to Technology

Conspicuousness, Obtrusiveness, and Obstinacy

In this technological era, Dasein uses technology to live its own life in a ready-to-hand manner which is regarded by Heidegger as an initial/primordial and a proper stage of investigating Dasein's existence because of its primordial experience of Being-in-the-(technological-)world. However, this primordial

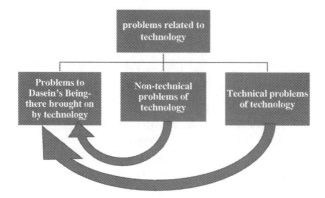

Figure 9.1 Justification of the problems to Dasein's existence posed by the problems related to technology.

manner of living in the world by using things as ready-to-hand does not last for long. The reasons for this are manifold.

The first reason is that the significant framework of referential totality of useful technological things would undergo a much more challenging task than that of normal non-technological things due to the frequent occurrences of domino effects;[9] that is, the sequence of the usage manner of "in-order-to-which" of technological things, to a very large extent, requires stronger necessary and sufficient conditions of using things—that is, without A (because of some technical problems), the Dasein cannot do B—in the usage. For instance, with the errors of my using my email, I cannot send an email to my colleague to ask him/her to fax a particular document to me, cannot do the follow-on tasks, and so on (unless I have another computer or a phone with emailing functions, but, normally speaking, a person would not have access to so much equipment at a time). On the contrary, the usage manner of "in-order-to-which" of non-technological things, relatively speaking, witnesses a lower frequency of domino effects by a series of *errors*, because Dasein can instantly pick up a substitute to replace the one with an *error* and tries to continue the following steps.

The second reason has something to do with the frequency of occurrences of technical problems/errors themselves happening to Dasein. Actually, the frequency of occurrences of technical problems of technology is higher than that of problems of non-technological things. Take a toothbrush versus an electronic toothbrush as an example. The former is used stably unless it wears out, with reduced usability, after being used for a long period of time, or it is broken because of Dasein's excessive strength when using it. Relatively speaking, this is easier to predict because Dasein can see the physical surface of a toothbrush at one glance. But this happening of problems of non-technological things is less frequent compared to that of an electronic toothbrush. It requires a battery,

Problems of Exacerbation to Dasein 137

which would run out from time to time, and it operates near watery areas which may trigger electronic problems. This is hard to predict because Dasein cannot see through the internal electronic structure of an electronic toothbrush and cannot predict when the water would finally affect the functions of the toothbrush. Therefore, from the perspective of predictability, Dasein tends to maintain the usability of a normal toothbrush because it can witness and better predict when it goes wrong, and thus the frequency of the occurrence of problems of a normal toothbrush is lower. On the contrary, as Dasein cannot see through the internal electronic structures at one glance, it cannot predict when the electronic toothbrush will go wrong, and it would care about it less. Therefore, due to less care from Dasein for the electronic toothbrush and the instability of electronic structures themselves, the frequency of the occurrence of problems of an electronic toothbrush is higher.

The third reason is that there may frequently be something unwelcomed and unexpected popping up to "stand in the way" of the use of technological things (Campbell 2019: 1663; Heidegger 1962: 103). Moreover, even though something unwelcomed and unexpected comes up to Dasein's devices, say computers, it does not mean that Dasein can be easy-going to let it be. Instead, it has to overcome the "something unwelcomed and unexpected" which "stands in its way" in order to go towards Dasein's ultimate goal. This description of "stand in the way" can be exemplified by looking at our use of computers where sometimes users' computers may be hacked, strange emails containing viruses arrive in users' e-mailboxes, and some unexpected websites come to make fake announcements to users. Of course, in these cases, users can simply ignore and cancel them. But for serious cases, users may have to spend much time mulling over possible solutions to address something unwelcomed and unexpected.

These three reasons address (1) the frequent occurrences of domino effects by a serious series of *errors*, (2) frequent occurrences of technical problems themselves happening to technological things, and (3) frequent occurrences of unexpected things that "stand in the way", when Dasein uses them. These three problems happening to technological things can be characterized as "the modes of conspicuousness, obtrusiveness and obstinacy which have the function of bringing to the fore the character of objective presence in what is at hand" (Heidegger 1996: 74). Among the three modes, obtrusiveness is the mode that can conspicuously address the first reason for the problems—domino effects by a series of *errors*. Obtrusiveness is a mode that characterizes Dasein's experience of this incomplete totality of (technological) things, as attested:

The more urgently we need what is missing, all the more obtrusive does that which is ready-to-hand become. … It reveals itself as something just present-at-hand and no more, which cannot be budged without the thing that is missing.

(Heidegger 1962: 103)

Obtrusiveness characterizes the powerlessness and disability of using things familiarly because of the lack of necessary links in the referential totality of things (Campbell 2019: 1663; ibid.). The lack of the necessary links can be referred to as the situation triggered by the domino effects by a series of *errors* where the initial breakdown use or any of the uses in the middle linkage causes the subsequent uses of the series to break down.

Among them, *conspicuousness* is a mode that characterizes technological things emerging from their readiness-to-hand as currently present-at-hand[10] and announcing themselves to Dasein's consciousness, addressing the frequent occurrences of technical problems themselves happening to technological things (Campbell 2019: 1662–63). Take the problem of the higher frequency of occurrences of technical problems/errors as an example. Frequently, Dasein tends to experience more technical problems/errors with the physical aspects of equipment, programming organizations, the arrangement of technical components, and so on, which would strip the technological things of their usefulness and their equipmentality.

Also, obstinacy is a mode characterizing the situation as something that is unwelcomed and unexpectedly "stands in the way" of our use of technological things (Campbell 2019: 1663; Heidegger 1962: 103). Obstinacy disturbs our progress towards the primary goal of the use of technological things and makes them present-to-hand, which "still lies before us and calls for our attending to it" (Campbell 2019: 1663; Heidegger 1962: 104).

The modes of conspicuousness, obtrusiveness, and obstinacy would stop Dasein from absorbing itself in the primordial world of using things in its everydayness and requires Dasein to have understanding of examinations of the problems/errors which made the technological things become malfunctioning, unfamiliar, and alien things (ibid.; Dreyfus 1991: 71). With a present-at-hand (not-primordial) manner, Dasein holds its Subject position to view the externalized Objects, technological things, as scientific/thematic things, in a Subject–Object dichotomous manner, which is what Heidegger is opposed to (Schmidt 2006: 51), as attested in:

> *such "Things" are encountered from out of a world in which they are ready-to-hand for Others—a world which is always mine too in advance.*
> (Heidegger 1962: 154)

This is because Heidegger thinks that Subject–Object dichotomous manner is not a proper living manner of Dasein to be Being-in-the-world with openness to the world primordially to subliminally interact with the world things (including technological things, of course) (Schmidt 2006: 51); this implies the "malfunction" of Dasein's absorption in everydayness. However, having said this, I do not mean that this present-at-hand horizon would last forever; instead, such a horizon would fade into a ready-to-hand manner, once errors and problems are fixed so that Dasein can adapt to what it has experienced through time. Even though the errors and problems cannot be fixed, after a certain period of time, the present-at-hand horizon would fade into oblivion because Dasein may have chosen an alternative for a way out inasmuch as inertia of using things matters most.

Lastly but equally importantly, one possible feature of the technological world can plausibly explain the frequent occurrences of these modes of Being of Things: conspicuousness, obstinacy, and obstructiveness. The feature is automation. As described by Christ (2015: 62), usually, technological things run autonomously according to the rules set by users (Dasein), and, due to the automated processes, the things are to be perceived as tools as Dasein would use with its physical hands (ibid.). As a result, because of the absence of human control, utilization, manipulation, observations, and examinations, the probability of technological things working in the modes of Being, conspicuousness, obstinacy, and obstructiveness tend to be greater (ibid.). Therefore, the point is to argue that not only do conspicuousness, obstinacy, and obstructiveness expose Dasein to the presence-at-hand of Things, the automation of the technological world exacerbates the exposure. From my standpoint, exacerbation of problems is more of a problem.

This section is intended to justify a part of Hypothesis (1): The former two problems related to technology are likely to become problems to Dasein's Being-there Brought on by Technology as true, by justifying the seriousness of a serious series of *errors* happening to technological Things which may affect Dasein's existence, Being-in-the-world.

Justification of the Problems to Dasein's Existence Posed by Non-technical Problems of Technology

Dasein lives in the technological word dealing with data and mathematical computing, "calculable quantities and relationships", figures, numerical units, statistical operations, and so on (Christ 2015: 58–63). As previously mentioned, when living with and dealing with these computing items, Dasein finds enjoyment and utilizes them well to overcome problems in a way better than Dasein does without technology. However, behind the scenes of the enjoyment and the utilization, when facing calculable quantities and mathematical data, Dasein is required to use its calculative thoughts and minds. This is human beings' mode of using Things that the ordinary may not find problematic. The reason for not being regarded as problematic is that these kinds of problems are non-technical, and they usually do not conspicuously and obviously draw Dasein's attention to recognizing them as obvious and repulsive "problems" as do *technical problems.*

This kind of non-technical problems are the *Problems and Dasein's Being-there Brought on by Technology*. The reason for this is that requiring Dasein to use its calculative thinking and mathematical skills means that Dasein needs to hold its Subject position and view sets of data and mathematical questions as a set of Objects in a present-at-hand manner, unlike readiness-to-hand in really average everydayness (which Heidegger criticizes). The Subject–Object dichotomously conscious experience happens due to the need for Dasein to treat sets of data and mathematical questions as tasks to solve (Schmidt 2006: 51). The act of "solving" implies that Dasein needs to treat them as Objects that need to be solved as students would tackle questions from their homework. However, the first reason is not what I would like to emphasize here. Instead, the second reason is such that the informationalised world exacerbates Dasein's exposure to experience the use of Things in a present-at-hand manner. To put the standpoint more mildly, the exacerbation of the problems is the *problem* that this paper would like address in the technological context. If so, then it is plausible to justify another part of Hypothesis (1): *The former two problems related to technology are likely to become Problems to Dasein's Being-there Brought on by Technology.*

Summary

From the technical problems of technology and non-technical problems of technology, we can see that the two of them help address and justify Hypothesis (1): *The former two problems related to technology are likely to become Problems to Dasein's Being-there Brought on by Technology.* Then the justification of this hypothesis can also help address the research questions of this paper:

1. How does the technological world trigger Dasein's present-at-hand understanding of the physical occurrences of technology more easily than the primitive life world?
2. Why is question 1 in opposition to the basic existential essence of Dasein?

Conclusion

This article addressed the problems—the exacerbation of presence-at-hand and inauthenticity in Dasein's way of living—in the setting of the modern technological era. There is a need to re-emphasize the importance of not regarding presence-at-hand as a serious problem according to Heidegger's philosophical theories, as this tends only to neutrally reflect Dasein's mode of Being and its relation to the world; however, the modern era of information technology would exacerbate the occurrence of Dasein's encounter with present-at-hand technological things and inauthenticity. That is, Dasein's relation to the worldhood of non-useful things as present-at-hand with the (temporary) breakdown of the referential totality of (technological) things, which exacerbates the frequency of occurrence of present-at-hand manners of using/treating Things.

Notes

1 Ontology and fundamental ontology: Heidegger is opposed to the so-called ontological task examining objective objects, and thus the "more" fundamental ontological task emerges as to questioning the Being of human beings, Dasein (Schmidt 2006: 52) (For Dasein, please refer to Note 3.)

2 Average everyday experience: a kind of experience that Heidegger emphasizes most with regard to our usual living experience which is a primordial experience before viewing any phenomenon of the Subject–Object dichotomy where we hold our own Subject position to view in opposition to an Object phenomenon outside of us, which is not what Heidegger emphasizes (Schmidt 2006: 51).
3 "Dasein" is used as a technical term designed by Heidegger himself to represent Being, literally translated as "Being there" from German to English (Anowai & Chukwujekwu 2019: 1; Schmidt 2006: 51); Being there is to be in the world, which is the most fundamental Dasein's mode of existence.
4 Dasein is being-possible as it itself is a set of possibilities by Being-in-the-World.
5 Understanding and Being-there: An existential structure that Dasein possesses and that Dasein must understand one way or another. It is primary in the primordial sense, meaning that understanding activities do not undergo thematic/theoretical examinations (Schmidt 2006: 63); instead, primordial understanding is the most fundamental understanding prior to all other, later thematic/theoretical understanding tasks to examine the "external" objects, which would be the present-at-hand understanding—a derivative of the primary primordial understanding (to be discussed later), as in "A based on B" pattern (refer to Heidegger 1996: 134). Such a primordial subliminal understanding "*constitutes the being of there in general*" (ibid.). Simply put in our plain language, the moment I understand, I find myself existing in the world (to be there).
6 Ready-to-hand: The mode of being of a thing; a thing shows itself to Dasein as something meaningful in the sense that the thing is revealed with reference to another thing and Dasein as well, the whole process of which Dasein primordially takes for granted with absorption in the use of the thing without reflecting on the referencing (Schmidt 2006: 65).
7 Contextual interpretation of Pragmatic Sense: Without the involvement of contexts, the way a thing is used is not performative and practical such that the way of use happens to be in relation to another thing to accomplish the end of the user (Recanati 2001). As is a speech act, the way a couple says "I do" to each other in the context of marriage is performative for speakers to accomplish the end of marriage, and therefore "I do" is not expressing the literal meaning but a hidden and intended meaning along with the context.
8 This is attested in: "In our 'description' of that environment which is closest to us – the work-world of the craftsman, for example, – the outcome was that along with the equipment to be found when one is at work [*in Arbeit*], those Others for whom the 'work' [*Werk*] is destined are 'encountered too'. If this is ready-to-hand, then there lies in the kind of Being which belongs to it (that is, in its involvement) an essential assignment or reference to possible wearers, for instance, for whom it should be 'cut to the figure'" (Heidegger 1962: 153).
9 Domino effect: The situation in which one event causes a series of related events, one following another (Cambridge Business English Dictionary @Cambridge University Press).
10 Present-at-hand: A derivative mode of a thing that is objectively present to Dasein (Schmidt 2006: 66). The reason for regarding it as derivative is that, as said previously, the most basic, primordial mode of a thing showing itself Dasein in the world-hood is readiness-to-hand, and this is the initial stage, upon which presence-at-hand of a thing becomes possible—i.e. derivative (Riemer & Johnston 2014: 277). Simply put, presence-at-hand is not a primordial, absorptive mode of a thing used but an examining mode of Subject–Object dichotomy.

References

Anowai, Eugene and Stephen Chukwujekwu (2019) "The Concept of Authentic and Inauthentic Existence in the Philosophy of Martin Heidegger: The 'Quarrel' of Communitarians and Libertarians", *Review of European Studies* 11, Canadian Center of Science and Education.Campbell, Jesse W. (2019) "Obtrusive, Obstinate and Conspicuous: Red Tape from a Heideggerian Perspective", *International Journal of Organizational Analysis* 27(5): 1657–72.

Christ, Oliver (2015) "Martin Heidegger's Notions of World and Technology in the Internet of Things Age", *Asian Journal of Computer and Information Systems* 3(2), 58–64.

Dreyfus, Hubert L. (1991) *Being-in-the-World: A Commentary on Heidegger's Being and Time*, Division I, Cambridge, Mass.: MIT Press.

Elpidorou, Andreas and Lauren Freeman (2015) "Affectivity in Heidegger I: Moods and Emotions in *Being and Time*", *Philosophy Compass* 10(10): 661–71.

Heidegger, Martin (1962) *Being and Time*, translated by John Macquarrie and Edward Robinson, New York: Harper & Row Publishers.

Heidegger, Martin (1996) *Being and Time*, translated by Joan Stambaugh, Albany: State of New York Press.

Heidegger, Martin (1999) *Ontology: Hermeneutics of Facticity*, translated by John van Buren, Bloomington, IN: Indiana University Press.

Lawhead, William F. (2002) *The Voyage of Discovery, A Historical Introduction to Philosophy* (2nd Ed.), Belmont, California: Wadsworth Publishing.

O'Brien, Mahon (2014) "Leaping Ahead of Heidegger: Subjectivity and Intersubjectivity in *Being and Time*", *International Journal of Philosophical Studies* 22(4): 534–51, DOI: 10.1080/09672559.2014.948719.

Recanati, Francois (2001) "What Is Said", *Synthese* 128(1/2): 75–91, retrieved from www.jstor.org/stable/20117147

Riemer, Kai and Robert B. Johnston (2014) "Rethinking the Place of the Artefact in Using Heidegger's Analysis of Equipment", *European Journal of Information Systems* 23(3): 273–88.

Schmidt, Lawrence Kennedy (2006) *Understanding Hermeneutics*, London and New York: Routledge.

Stefani, Jaqueline and N.O. da Cruz (2019) "Understanding and Language in Heidegger: Ex-sistence, Ontological Openness and Hermeneutics", *Bakhtiniana: Revista de Estudos do Discurso* 14(2): 112–27.

10 Ethically Speaking

Opportunities and Risks of AI Chatbots Showing Empathy to Customers during Service Encounters

Yeung Wing Lok

Introduction

Recent advances in artificial intelligence (AI) have brought considerable progress in areas such as computer vision and speech recognition, thereby enabling computers and robots to perform an increasing number of human tasks and outperform humans in many cases (Dargan et al. 2020). However, there have also been cases where the performance of AI is called into question for ethical reasons such as fairness and equal treatment (Chamorro-Premuzic et al. 2019; Chouldechova and Roth, 2020; Simons et al. 2021). On the other hand, people tend to hold different expectations for the performance of AI as opposed to a human and oftentimes do not apply the same ethical standards (Banks 2020). For instance, experimental studies have shown that people would apply different moral standards to humans vs. robots on their decisions made during a dilemma situation (Malle et al. 2015; Voiklis et al. 2016). Recognising such differences is necessary for the proper design of AI applications in ethical terms.

This chapter focuses a specific type of AI application, namely customer service chatbots, and examines a particular ethical requirement for its design. Chatbots are computer programs that simulate the verbal behaviour of an individual human conversing with other humans through the medium of text only. Chatbots have been around since 1960s (Weizenbaum 1966) and have become increasingly popular with smartphone users in recent years (Dale 2016). Early chatbots relied on using pattern-matching rules to parse human inputs and to produce scripted responses in limited domains (Sankar et al. 2008). Recent advances in corpus-based natural language processing (NLP) AI techniques have further enhanced chatbots' abilities to conduct dialogues in a more human-like fashion (Peng and Ma 2019).

Meanwhile, the widespread use of smartphone-based messaging and social media apps has contributed to the viability of chatbots as a means of providing customer service by businesses. Chatbots have been adopted by companies to serve their clients in e-commerce (Qiu et al. 2017), health care (Huang et al. 2018), banking (Shen 2018), travel (Argal et al. 2018), etc. Predictions

DOI: 10.4324/9781003376491-11

suggest that 75–90% of customer service interactions in the banking and health care industries will be handled by chatbots by 2022 (Han 2021).

Customers enjoy using chatbots for various reasons such as efficiency and convenience (Brandtzaeg and Følstad 2017). In particular, studies have shown that being sensitive to users' emotions and showing empathy are desirable abilities of chatbots (Chaves and Gerosa 2021; Rapp et al. 2021). However, endowing computers and robots with emotions and empathy has also raised some ethical concerns with regard to their potential influence on human decision making in, e.g. purchase decisions, privacy disclosure, etc. (High-Level Expert Group on Artificial Intelligence 2019; Rapp et al. 2021).

This chapter reviews a number of ethical issues on the showing of empathy in the context of customer service. Specifically:

1. Why, ethically speaking, should companies show empathy to customers during service encounters?
2. Why should chatbots do the same, despite not being human?
3. What is wrong with showing empathy that contradicts one's internal feelings?
4. Why should chatbots avoid showing empathy that involves their internal feelings?
5. What are some potential harms of chatbots showing empathy to customers?

To answer these questions, theoretical ideas and empirical evidence were drawn from relevant literature in ethics, psychology, human computer interaction, marketing and customer service, chatbot design, and more.

In the following, we first review why customers can feel unfairly treated if agents do not show empathy during their service encounters. Then we consider whether the same can happen when chatbots substitute for human agents. Finally, we consider some ways in which chatbot empathy could be misused and cause harm to customers. The last part is the conclusion of this chapter.

Empathy in Customer Service

Empathy as a human trait or ability has been given various definitions in different contexts (Snow 2000). In the context of customer service, it has been considered as one's ability to sense and react to customers' thoughts, feelings, and experiences during service encounters (Wieseke et al. 2012). According to Barrett-Lennard (1981), the process of empathy involves a cycle of three phases: (1) empathic understanding (an agent senses the feelings expressed by a customer), (2) empathic expression (the agent shows his/her understanding of the customer's feelings in response), and (3) empathic reception (the customer recognises that his/her own feelings are shared with the agent). After phase 3, the process can continue from phase 1 with the customer expressing more about his/her experience and feelings in conjunction with the agent's empathic feedback.

Empathy is generally regarded as a desirable employee competence in customer service (D'Cruz and Noronha 2008; Deery and Kinnie 2002; Frenkel et al. 1998; Clark et al. 2013). It is considered one of the key factors affecting customers' perceptions of service quality, according to the SERVQUAL measurement scale (Parasuraman et al. 1988). Researchers have found positive effects of employee empathy on customer perceptions of service quality (Groth and Grandey 2012), customer satisfaction (Wieseke et al. 2012), and business performance (Ye et al. 2017). The ability to understand customers' experiences, to take their perspectives, and to connect with customers emotionally can help increase satisfaction (Simon 2013) and gain trust from them (Aggarwal et al. 2005; Simon 2013).

Unfairness of Not Showing Empathy to Customers

Hodroff (2014) considered customer service from the point of view of business ethics and asserted that customers are entitled to customer service that is genuine and fair. Anything done by a business that makes it more difficult to resolve issues with ease or satisfaction through customer service is an ethical issue (Hodroff 2014). In particular, customer dissatisfaction can result from employees' poor responses to customers' requests for assistance, other than from the issues that have prompted the requests (Bitner et al. 1990).

For instance, if an employee senses that a customer feels offended by something just said (unintentionally), an immediate clarification or a quick apology could relieve the customer from a feeling of being wrongly or unfairly treated; failing that, the employee could be seen as unethical. In situations like this, the customer's emotional expression allows the employee to understand and respond in an appropriate way, i.e. to show empathy, so as to keep their relationship on a good course (Kennedy-Moore and Watson 2001). When customers are shown empathy during service encounters, they feel that they are not just being understood for what they have said but also how they feel.

According to the ideas of service fairness (Seiders and Berry 1998; Tax et al. 1998), fair service treatment involves empathy and other attributes such as politeness and honesty, which are considered essential as a matter of interactional justice and fair treatment (Goodwin and Ross 1992). Schneider and Bowen (1999) considered service fairness or justice a core customer need that must be satisfied to avoid customer outrage. Similar to being rude, treating a customer unfairly is commonly regarded as unethical behaviour during service encounters (Schwepker Jr and Hartline 2005).

In short, it is considered unfair and hence unethical if companies do not show empathy to customers during service encounters.

When Empathy Is Fabricated

Empathy requires attentiveness (Parasuraman et al. 1991). A grounded study of call centre operations (Clark et al. 2013) found that empathic expressions contributed to effective customer service only when agents listened attentively to assess any needs for empathy and provided necessary communicative responses to meet those needs expeditiously. However, attentiveness can be "faked", for instance, by simply repeating customers' words, and customers can tell.

Customers may wonder whether the employees are being sincere (Clark et al. 2013) or simply acting out from scripts to induce customer satisfaction and trust (Aggarwal et al. 2005). From a sociolinguistic point of view, expressive speech acts (such as "Thanking") without sincerity (e.g. feeling grateful) can be associated with attempting to deceive (Mann and Kreutel 2004). Ford (1996) regarded the fabrication of images not reflecting employees' true feelings, behaviours, or attitudes as a form of deception. This is also known in the literature as emotional labour (Gray 2000) or emotional management (Hochschild 1979), i.e. attempts by a service agent to show customers' emotions which are inconsistent with his/her true feelings. It could be, for instance, that the agent is in a bad mood or has been annoyed by the previous customer. Emotional labour has also been associated with ethical issues related to stress and burnouts in the workplace (Hochschild 2012).

Chatbot Empathy

Do customer service chatbots need to show empathy during service encounters? Empirical studies (James et al. 2018, 2020) found that, if emotions are added to words in robots' speech, humans can perceive empathy and prefer it over robotic-sounding voice. Chatbots that gave empathic responses were rated higher by users in a number of studies (Daher et al. 2020; Hu et al. 2018; Xu et al. 2017; Liu and Sundar 2018). Baron (2015) suggested, from a linguistic-functional point of view, that people should welcome empathic expressions when talking to either humans or robots.

According to the CASA (computers are social actors) paradigm (Nass et al. 1993), people are prone to apply social rules (e.g. be courteous) and norms (e.g. praise from others are perceived as more valid than self-praise) in their perception of and interactions with computers which exhibit certain social cues (e.g. speaking in a human-like friendly voice). They do so without mistaking in any way computers as humans, and they would also recognise their social ways of reacting/responding to computers as unnecessary and inappropriate. Liu and Sundar (2018) found support for the CASA paradigm regarding the social effects of chatbot empathy in their experiments on users' perception of a medical advice chatbot. Given that the showing of empathy in customer service is generally an ethical social norm, chatbots that show no empathy to customers may well be regarded as violating this norm and be considered as unethical. Miller et al. (2015) argued that as human users became reliant on human-like robots, they could ignore the latter's non-human aspects even in the presence of reminders and attribute human agency and moral accountability to them.

Emotional Expression and Authenticity

As previously mentioned, customers can feel uneasy with a human agent giving fabricated or inauthentic empathic expressions that do not reflect the agent's true feelings. A recent user study (Mariacher et al. 2021) found that users would not take a chatbot's empathy seriously if the latter was perceived as inauthentic. Assuming that empathy is morally important also for customer service chatbots, how could chatbots avoid making empathic expressions that are perceived as inauthentic by customers?

According to a recent study (Seitz and Bekmeier-Feuerhahn 2021), users tend to dislike chatbots giving empathic expressions such as "I am sorry …" with references to self-emotions (e.g. remorse) which are not meant to exist. A plausible reason is that users find the expressions less authentic (Seitz and Bekmeier-Feuerhahn 2021). According to the Searle's Speech Act Theory (Searle 1969), the sincerity of an empathic expression can be viewed as dependent on the sincerity condition of the speech act type used in the expression. If an empathic expression coming from a chatbot uses a speech act type (e.g. "Thank", "Congratulate") that has a sincerity condition based on the chatbot's emotion (e.g. grateful, pleased), it is prone to be perceived as inauthentic.

For instance, instead of saying, "I am sorry for the delay …", which suggests feeling remorse and may well sound inauthentic from a chatbot, using a more neutral expression like "Let me help you with the delay …" would be a better choice. Another suggestion is that being honest about its artificial nature can help a chatbot be perceived as more authentic (Mariacher et al. 2021).

Anthropomorphism and Ethical Risks of Chatbot Empathy

Studies have shown that customer service chatbots designed with human-likeness can increase anthropomorphism in customers (Araujo 2018). Anthropomorphism has been regarded by some researchers as the attribution of human form, a human-like mind, or mental states to a non-human entity (Waytz et al. 2010). Schroeder and Epley (2016) pinpointed the attribution of a human-like mind to a machine as the essence of the anthropomorphism of computers. Gray et al. (2007) identified a range of mental capacities that people would attribute to a non-human entity when perceiving a human mind in it. Some of these capacities are agency-related such as self-control, morality, and emotion recognition; others are experience-related, such as hunger, fear, and pain. This may help explain how chatbot empathy can be conducive to anthropomorphism, since empathy requires emotion recognition and showing empathy to customers represents a moral act for customer service chatbots.

There has been much discussion on the ethical risks of anthropomorphising computers and robots (Leong and Selinger 2019; Sharkey and Sharkey 2021). Those relevant to customer service chatbots are discussed here.

Empathy as a Mask

The showing of empathy has been associated with gaining customers' trust (Aggarwal et al. 2005; Simon 2013). In their experiment, Klein et al. (2002) showed that chatting agents could alleviate frustration in computer users through active listening, empathy, and sympathy. However, Picard and Klein (2002) pointed out that repeated use of this strategy without solving users' actual problems would be deemed disingenuous or manipulative. If chatbot empathy is used as a manipulative device to exploit customers' emotions or to gain their trust as a way to mask the company's failure or reluctance to provide certain services or fix issues, that becomes unethical (Jiang et al. 2016).

Unintended Behavioural Influence

Studies have shown that chatbot anthropomorphism can influence consumers' attitudes, emotional connection, and satisfaction with companies (Araujo 2018). A survey study (Han 2021) showed that participants could perceive anthropomorphism and social presence with chatbots and were inclined to use them to purchase products in the future. Empathy has also been exploited by robots in eliciting other prosocial behaviours such as helpfulness from humans (Gonsior et al. 2012). However, exploiting chatbot anthropomorphism also bears risks of unintended negative influence on human behaviours. Liu and Sundar (2018) outlined a taxonomy of negative impacts of anthropomorphism. For instance, customers may get worried about hurting a chatbot's "feelings" or offending it with inappropriate words.

Illusion of Care

Some ethicists argue against computers and robots expressing emotions to humans because the latter could misbelieve that machines care for them (Sharkey and Sharkey 2021) or could form unidirectional emotional bonds with the machines (Scheutz 2011). Anthropomorphised agents can act as powerful agents of social connection particularly for individuals with a strong desire for connection (Epley et al. 2007). We can imagine customers of certain dispositions such as loneliness being more vulnerable to the illusion of care from chatbots, and this could render them even less inclined to seek support from real humans, thus reinforcing their loneliness.

Conclusion

Advances in AI have brought new opportunities for companies to provide customer service on popular online chat platforms using chatbots. This chapter explains why designing chatbots to show empathy is not only desirable but necessary from an ethical point of view. Customers would feel being treated unfairly if their emotions do not get adequate recognition and responses from service chatbots. On the other hand, empathic expressions with references to chatbots' "feelings" would sound inauthentic to customers. Finally, since chatbot empathy may cause anthropomorphism in users, there are some risks of chatbots being misused against customers' interest, as well as customers mistaking the existence of social connection and care from chatbots.

References

Aggarwal, Praveen, Stephen B. Castleberry, Rick Ridnour, and C. David Shepherd (2005) "Salesperson Empathy and Listening: Impact on Relationship Outcomes", *Journal of Marketing Theory and Practice* 13(3): 16–31.

Araujo, Theo (2018) "Living up to the Chatbot Hype: The Influence of Anthropomorphic Design Cues and Communicative Agency Framing on Conversational Agent and Company Perceptions", *Computers in Human Behavior* 85: 183–89.

Argal, A., Siddharth Gupta, Ajay Modi, Pratik Pandey, Simon Shim, and Chang Choo (2018) "Intelligent Travel Chatbot for Predictive Recommendation in Echo Platform", in *2018 IEEE 8th Annual Computing and Communication Workshop and Conference (CCWC)*, IEEE, 176–83.

Banks, Jaime (2020) "Good Robots, Bad Robots: Morally Valanced Behavior Effects on Perceived Mind, Morality, and Trust", *International Journal of Social Robotics* 13: 2021–38.

Baron, Naomi S. (2015) "Shall We Talk? Conversing with Humans and Robots", *The Information Society* 31(3): 257–64.

Barrett-Lennard, Godfrey T. (1981) "The Empathy Cycle: Refinement of a Nuclear Concept", *Journal of Counseling Psychology* 28(2): 91–100.

Bitner, Mary Jo, Bernard H. Booms, and Mary Stanfield Tetreault (1990) "The Service Encounter: Diagnosing Favorable and Unfavorable Incidents", *Journal of Marketing* 54(1): 71–84.

Brandtzaeg, Petter Bae and Asbjørn Følstad (2017) "Why People Use Chatbots", in *International Conference on Internet Science*, Springer, 377–92.

Chamorro-Premuzic, Tomas, Frida Polli, and Ben Dattner (2019) "Building Ethical AI for Talent Management", *Harvard Business Review* 21(November): 1–15.

Chaves, Ana Paula and Marco Aurelio Gerosa (2021) "How Should My Chatbot Interact? A Survey on Social Characteristics in Human–chatbot Interaction Design", *International Journal of Human–Computer Interaction* 37(8): 729–58.

Chouldechova, Alexandra and Aaron Roth (2020) "A Snapshot of the Frontiers of Fairness in Machine Learning", *Communications of the ACM* 63(5): 82–89.

Clark, Colin Mackinnon, Ulrike Marianne Murfett, Priscilla S. Rogers, and Soon Ang (2013) "Is Empathy Effective for Customer Service? Evidence from Call Center Interactions", *Journal of Business and Technical Communication* 27(2): 123–53.

Daher, Karl, Jacky Casas, Omar Abou Khaled, and Elena Mugellini (2020) "Empathic Chatbot Response for Medical Assistance", in *Proceedings of the 20th ACM International Conference on Intelligent Virtual Agents*, 1–3.

Dale, Robert (2016) "The Return of the Chatbots", *Natural Language Engineering* 22(5): 811–17.

Dargan, Shaveta, Munish Kumar, Maruthi Rohit Ayyagari, and Gulshan Kumar (2020) "A Survey of Deep Learning and its Applications: A New Paradigm to Machine Learning", *Archives of Computational Methods in Engineering* 27(4): 1071–92.

D'Cruz, Premilla and Ernesto Noronha (2008) "Doing Emotional Labour: The Experiences of Indian Call Centre Agents", *Global Business Review* 9(1): 131–47.

Deery, Stephen and Nicholas Kinnie (2002) "Call Centres and beyond: A Thematic Evaluation", *Human Resource Management Journal* 12(4): 3–13.

Epley, Nicholas, Adam Waytz, and John T. Cacioppo (2007) "On Seeing Human: A Three- factor Theory of Anthropomorphism", *Psychological Review* 114(4): 864–86.

Ford, Wendy S. Zabava (1996) "Ethics in Customer Service: Critical Review and Research Agenda", *Electronic Journal of Communication/La Revue Electronique de Communication* 6(4).

Frenkel, Stephen J., May Tam, Marek Korczynski, and Karen Shire (1998) "Beyond Bureaucracy? Work Organization in Call Centres", *International Journal of Human Resource Management* 9(6): 957–79.

Gonsior, Barbara, Stefan Sosnowski, Malte Buß, Dirk Wollherr, and Kolja Kühnlenz (2012) "An Emotional Adaption Approach to Increase Helpfulness towards a Robot", in *2012 IEEE/RSJ International Conference on Intelligent Robots and Systems*, IEEE, 2429–36.

Goodwin, Cathy and Ivan Ross (1992) "Consumer Responses to Service Failures: Influence of Procedural and Interactional Fairness Perceptions", *Journal of Business Research* 25(2): 149–63.

Gray, Heather M., Kurt Gray, and Daniel M. Wegner (2007) "Dimensions of Mind Perception", *Science* 315(5812): 619.

Gray, Wayne D. (2000) "The Nature and Processing of Errors in Interactive Behavior", *Cognitive Science* 24(2): 205–48.

Groth, Markus and Alicia Grandey (2012) "From Bad to Worse: Negative Exchange Spirals in Employee–customer Service Interactions", *Organizational Psychology Review* 2(3): 208–33.

Han, Min Chung (2021) "The Impact of Anthropomorphism on Consumers' Purchase Decision in Chatbot Commerce", *Journal of Internet Commerce* 20(1): 46–65.

High-Level Expert Group on Artificial Intelligence (2019) "Ethics Guidelines for Trustworthy AI", available at https://ec.europa.eu/digital-single-market/en/news/ethics-guidelines-trustworthy-ai, retrieved on 18 January 2021.

Hochschild, Arlie Russell (1979) "Emotion Work, Feeling Rules, and Social Structure", *American Journal of Sociology* 85(3): 551–75.

Hochschild, Arlie Russell (2012) *The Managed Heart: Commercialization of Human Feeling*, Oakland, California: University of California Press.

Hodroff, Matthew (2014) "Customer Service Is a Business Ethics Issue", *Ethikos* September / October, available at SSRN: https://ssrn.com/abstract=2494475

Hu, Tianran, Anbang Xu, Zhe Liu, Quanzeng You, Yufan Guo, Vibha Sinha, Jiebo Luo, and Rama Akkiraju (2018) "Touch Your Heart: A Tone-aware Chatbot for Customer Care on Social Media", in *Proceedings of the 2018 CHI Conference on Human Factors in Computing Systems*, 1–12.

Huang, Chin-Yuan, Ming-Chin Yang, Chin-Yu Huang, Yu-Jui Chen, Meng-Lin Wu, and Kai-Wen Chen (2018) "A Chatbot-supported Smart Wireless Interactive Healthcare System for Weight Control and Health Promotion", in *2018 IEEE International Conference on Industrial Engineering and Engineering Management (IEEM)*, IEEE, 1791–95.

James, Jesin, B. T. Balamurali, Catherine I. Watson, and Bruce MacDonald (2020) "Empathetic Speech Synthesis and Testing for Healthcare Robots", *International Journal of Social Robotics* 13: 2119–37.

James, Jesin, Catherine Inez Watson, and Bruce MacDonald (2018) "Artificial Empathy in Social Robots: An Analysis of Emotions in Speech", in *2018 27th IEEE International Symposium on Robot and Human Interactive Communication (RO-MAN)*, IEEE, 632–37.

Jiang, Kaifeng, Jia Hu, Ying Hong, Hui Liao, and Songbo Liu (2016) "Do it Well and Do it Right: The Impact of Service Climate and Ethical Climate on Business Performance and the Boundary Conditions", *Journal of Applied Psychology* 101(11): 1553.

Kennedy-Moore, Eileen and Jeanne C. Watson (2001) "How and When Does Emotional Expression Help? *Review of General Psychology* 5(3): 187–212.

Klein, Jonathan, Youngme Moon, and Rosalind W. Picard (2002) "This Computer Responds to User Frustration: Theory, Design, and Results", *Interacting with Computers* 14(2): 119–40.

Leong, Brenda and Evan Selinger (2019) "Robot Eyes Wide Shut: Understanding Dishonest Anthropomorphism", in *Proceedings of the Conference on Fairness, Accountability, and Transparency*, 299–308.

Liu, Bingjie and S. Shyam Sundar (2018) "Should Machines Express Sympathy and Empathy? Experiments with a Health Advice Chatbot", *Cyberpsychology, Behavior, and Social Networking* 21(10): 625–36.

Malle, Bertram F., Matthias Scheutz, Thomas Arnold, John Voiklis, and Corey Cusimano (2015) "Sacrifice One for the Good of Many? People Apply Different Moral Norms to Human and Robot Agents", in *2015 10th ACM/IEEE International Conference on Human-Robot Interaction (HRI)*, IEEE, 117–24.

Mann, William C. and Jörn Kreutel (2004) "Speech Acts and Recognition of Insincerity", in *Proceedings of the Eighth Workshop on the Semantics and Pragmatics of Dialogue*, 64–68.

Mariacher, Natascha, Stephan Schlögl, and Alexander Monz (2021) "Investigating Perceptions of Social Intelligence in Simulated Human-chatbot Interactions", in *Progresses in Artificial Intelligence and Neural Systems*, Springer, 513–29.

Miller, Keith, Marty J. Wolf, and Frances Grodzinsky (2015) "Behind the Mask: Machine Morality", *Journal of Experimental and Theoretical Artificial Intelligence* 27(1): 99–107.

Nass, Clifford, Jonathan Steuer, Ellen Tauber, and Heidi Reeder (1993) "Anthropomorphism, Agency, and Ethopoeia: Computers as Social Actors", in *Interact'93 and CHI'93 Conference Companion on Human Factors in Computing Systems*, 111–12.

Parasuraman, Arun, Leonard L. Berry, and Valarie A. Zeithaml (1991) "Understanding Customer Expectations of Service", *Sloan Management Review* 32(3): 39–48.

Parasuraman, Arun, Valarie A. Zeithaml, and Leonard L. Berry (1988) "Servqual: A Multiple-item Scale for Measuring Consumer Perception of Service Quality", *Journal of Retailing* 64(1): 12–40.

Peng, Zhenhui and Xiaojuan Ma (2019) "A Survey on Construction and Enhancement Methods in Service Chatbots Design", *CCF Transactions on Pervasive Computing and Interaction* 1(3): 204–23.

Picard, Rosalind W. and Jonathan Klein (2002) "Computers That Recognise and Respond to User Emotion: Theoretical and Practical Implications", *Interacting with Computers* 14(2): 141–69.

Qiu, Minghui, Feng-Lin Li, Siyu Wang, Xing Gao, Yan Chen, Weipeng Zhao, Haiqing Chen, Jun Huang, and Wei Chu (2017) "Alime Chat: A Sequence to Sequence and

Rerank Based Chatbot Engine", in *Proceedings of the 55th Annual Meeting of the Association for Computational Linguistics* (Volume 2: Short Papers), 498–503.

Rapp, Amon, Lorenzo Curti, and Arianna Boldi (2021) "The Human Side of Human-chatbot Interaction: A Systematic Literature Review of Ten Years of Research on Text-based Chatbots" *International Journal of Human-Computer Studies* 151: 102630.

Sankar, Gopal Ravi, Jean Greyling, Dieter Vogts, and Mathys C. du Plessis (2008) "Models towards a Hybrid Conversational Agent for Contact Centres", in *Proceedings of the 2008 Annual Research Conference of the South African Institute of Computer Scientists and Information Technologists on IT Research in Developing Countries: Riding the Wave of Technology, SAICSIT '08*, New York, the United States, Association for Computing Machinery, 200–9.

Scheutz, Matthias (2011) "The Inherent Dangers of Unidirectional Emotional Bonds between Humans and Social Robots", in Patrick Lin, Keith Abney, and George A. Bekey (eds.) *Robot Ethics: The Ethical and Social Implications of Robotics*, Cambridge, Mass.: MIT Press, 205–22.

Schneider, Benjamin and David E. Bowen (1999) "Understanding Customer Delight and Outrage", *Sloan Management Review* 41(1): 35–45.

Schroeder, Juliana and Nicholas Epley (2016) "Mistaking Minds and Machines: How Speech Affects Dehumanization and Anthropomorphism", *Journal of Experimental Psychology: General* 145(11): 1427–37.

Schwepker Jr, Charles H. and Michael D. Hartline (2005) "Managing the Ethical Climate of Customer-contact Service Employees", *Journal of Service Research* 7(4): 377–97.

Searle, John R. (1969) *Speech Acts: An Essay in the Philosophy of Language*, volume 626, Cambridge: Cambridge University Press.

Seiders, Kathleen and Leonard L. Berry (1998) "Service Fairness: What it Is and Why it Matters", *Academy of Management Perspectives* 12(2): 8–20.

Seitz, Lennart and Sigrid Bekmeier-Feuerhahn (2021) "Empathic Healthcare Chatbots: Comparing the Effects of Emotional Expression and Caring Behavior", in *Proceedings of ICIS 2021*, Austin, Texas.

Sharkey, Amanda and Noel Sharkey (2021) "We Need to Talk about Deception in Social Robotics!" *Ethics and Information Technology* 23: 309–16.

Shen, Alice (2018) "HSBC's Amy and Other Soon-to-be Released AI Chatbots Are about to Change the Way We Bank", *South China Morning Post*.

Simon, Françoise (2013) "The Influence of Empathy in Complaint Handling: Evidence of Gratitudinal and Transactional Routes to Loyalty", *Journal of Retailing and Consumer Services* 20(6): 599–608.

Simons, Joshua, Sophia Adams Bhatti, and Adrian Weller (2021) "Machine Learning and the Meaning of Equal Treatment", in *Proceedings of the 2021 AAAI/ACM Conference on AI, Ethics, and Society*, 956–66.

Snow, Nancy E. (2000) "Empathy", *American Philosophical Quarterly* 37(1): 65–78.

Tax, Stephen S., Stephen W. Brown, and Murali Chandrashekaran (1998) "Customer Evaluations of Service Complaint Experiences: Implications for Relationship Marketing", *Journal of Marketing* 62(2): 60–76.

Voiklis, John, Boyoung Kim, Corey Cusimano, and Bertram F. Malle (2016) "Moral Judgments of Human vs. Robot Agents", in *2016 25th IEEE International Symposium on Robot and Human Interactive Communication (RO-MAN)*, IEEE, 775–80.

Waytz, Adam, John Cacioppo, and Nicholas Epley (2010) "Who Sees Human? The Stability and Importance of Individual Differences in Anthropomorphism", *Perspectives on Psychological Science* 5(3): 219–32.
Weizenbaum, Joseph (1966) "ELIZA—A Computer Program for the Study of Natural Language Communication between Man and Machine", *Communications of the ACM*, 9(1): 36–45.
Wieseke, Jan, Anja Geigenmüller, and Florian Kraus (2012) "On the Role of Empathy in Customer-employee Interactions", *Journal of Service Research* 15(3): 316–31.
Xu, Anbang, Zhe Liu, Yufan Guo, Vibha Sinha, and Rama Akkiraju (2017) "A New Chatbot for Customer Service on Social Media", in *Proceedings of the 2017 CHI Conference on Human Factors in Computing Systems*, 3506–10.
Ye, Jun, Beibei Dong, and Ju-Yeon Lee (2017) "The Long-term Impact of Service Empathy and Responsiveness on Customer Satisfaction and Profitability: A Longitudinal Investigation in a Healthcare Context", *Marketing Letters* 28(4): 551–64.

Index

A Dictionary of Translation Technology 44
Adorno, Theodor W. 69
AI chatbot 143
Amity Drama Club 19, 22
Artificial Intelligence (AI) 55–6, 66, 71, 143
A Thousand Plateaus 29
Auden, W. H. 23
Auslander, Phillip 26

Baker, Mona 76
Beecham, Harriet 21
Being and Time 127
Big Data 4, 10, 52, 55, 71
Borgmann, Albert 72
Bowker, Lynne 44
Building a Character 21

camera translation 63
Candide project 51
Chan, Chi-wah 23
Chan, Sin-wai 44
chatbot 143–4, 146–9; chatbot anthropomorphism 148–9
Claudius 22
clean-data statistical machine translation 52
cloud-based computer-aided translation system 58
Coetzee, J. M. 30–1, 33–6, 38, 40
collaborative translation 58
common-sense knowledge database 49–50
computer-aided human translation 45, 48
computer-aided translation 44–7, 56, 59–60, 62
contextualized translation 53, 55

corpus-based translation studies (CBTS) 76
Corpus of Contemporary American English (COCA) 80
crowdsourcing 60, 63

Dasein 127–40
data 4–5; firsthand data 5
Deleuze, Gilles 29
dialogue interpreting 47–8
digital humanities 4
direct machine translation (DMT) 48
direct mapping 49
direct matching 49
dirty-data statistical machine translation 52
domain-specific statistical machine translation 52
draft translation 48
Du Fu 25
Du Fu 2.0 25

Epstein, Mikhail 28
exact translation 51
example-based machine translation (EBMT) 48, 50

FAHQT 48
Firth, J. R. 31
Fisher-Licht, Erika 21
Fu Yueh Mei 22

general applications statistical machine translation 52
Google 52
Google Translate 48, 53, 60–1
Google Translator Toolkit 60
Grim, Patrick 66
Guattari, Felix 29

Hamilton, Grant 28
Hamlet/Hamlet 22
Harari, Yuval Noah 71
Heidegger, Martin 69, 127, 130, 135, 137–8, 140
Homo Deus 71
Hong Kong Repertory Company 19, 21, 23
Horizon Theatre Group 22
human-aided computer translation 45, 48
human interpreting 45
human translation 45–6, 58
Hunt, Timothy 62

IBM TJ Watson Research Center 51
ICQ 24
image translation 63
indirect data 5
information translation 47–8
interlingual matching 50
Internet-based computer-aided translation system 58
interpreting 46–7
In the Wire 19
intralingual matching 50

Jockers, Matthew 29
Johnson, Steve 24

Kapp, Ernst 67
knowledge-based machine translation (KBMT) 48–9
knowledge-based system 50

Lam, Shu Yan 93
Lam, Yee Man 93
Lau, Hok-Yin 127
literary translation 55
Littlewood, Joan 21
Liu, Jianwen 76
Liu, Kanglong 76
LIVAC 4, 15–16
localization 44–7, 59; full localization 60; partial localization 60
Luk, Yun-tong Thomas 19

machine translation (MT) 44–9, 52–5, 59–60
Mak, Kelly 4
Marcuse, Herbert 69
Marx, Karl 69
metaverse 24, 26 113, 117, 122

Microsoft 52
Microsoft Translator 53
Midsummer Night's Dream 25
Miss Saigon 19
mobile translation 62–3
Modern Drama 21
Mok, Kenny 4
Moretti, Franco 29–30
Mou, Zongsan 67

Nagao, Makoto 50
neural machine translation 48, 52–4
New York International Fringe Festival 19
Ng, Kwan-kwan 93
Nuremberg Trials 46

official interpreting 46
Oh, What Lovely War 21
Olivier, Laurence 22

partial translation 46
philosophy-aided technology 66–7, 73
probable translation 51
Prospect Theatre 25
Pym, Anthony 59

Quayon, Ato 37

Rape Virus 23
reverse localization 59–60
Royal Shakespeare Company 25
rule-based machine translation (RBMT) 48, 50–1

SDL-Trados 62
Septuagint (Hebrew Bible into Greek by Seventy Greek Scholars) 46
Shakespeare, William 22
Shatin Theatre Company 22
Shen, Jiandong 113
simultaneous interpreting 46–7
Social networking service (SNS) 70–2
software localization 45, 47, 59
specialized knowledge database 49–50
Speech Act Theory 147
speech translation 44–8, 60–1
Stanislavsky, Konstantin 21
statistical machine translation (SMT) 48, 51–2
Sunshine, the Station Master 22
Su, Yanfang 76
Systran 52–3

techno-humanities 28, 66
Techno-Humanities Research Centre 66
technological determinism 72
technology-aided philosophy 66–7
techno-philosophy 66
textual translation 46–7
TF-IDF 31, 33, 35, 38
The Future of Translation Technology: Towards a World without Babel 44
The Search for the Spring Willow 20
Tragedy and Postcolonial Literature 37
translation memory database 58
translation technology 44–8, 62
TranTech 44–5, 62
TR Corpus 80
Tsou, Benjamin K. 4

Uncle Tom's Cabin 21
Underwood, Ted 29

visual translation 63

Waiting for the Barbarians 30–40
Webpage localization 45, 47, 59
Williams, Tennessee 24
Wong, Pui Yan 113
Wong, Wai Chung 113
Wooster Group, The 19
word-by-word machine translation system 49
WWW.com 23

Xiong, Yuan Wei 22

Yeung, Wing Lok 143
Ying, Koon Kau 66
Yu, Jung-jun 23

Zuni Icosahedron 19

Printed in the United States
by Baker & Taylor Publisher Services